PHalarope Books

PHalarope Books are designed specifically for the amateur naturalist. These volumes represent excellence in natural history publishing. Most books in the PHalarope series are based on a nature course or program at the college or adult education level or are sponsored by a museum or nature center. Each PHalarope book reflects the author's teaching ability as well as writing ability. Among the books:

At the Sea's Edge: An Introduction to Coastal Oceanography for the Amateur Naturalist
William T. Fox/illustrated by Clare Walker Leslie

Exploring Tropical Isles and Seas: An Introduction for the Traveler and Amateur Naturalist
Frederic Martini

Pond and Brook: A Guide to Nature Study in Freshwater Environments
Michael J. Caduto

The Seaside Naturalist: A Guide to Nature Study at the Seashore
Deborah A. Coulombe

Sharks: An Introduction for the Amateur Naturalist
Sanford A. Moss

The following oceanography texts are also published by Prentice-Hall:

An Introduction to Oceanography, 3d Edition
David A. Ross

The World Ocean: An Introduction to Oceanography, 2nd Edition
William A. Anikouchine and Richard W. Sternberg

Oceanography: A View of the Earth, 3rd Edition
M. Grant Gross

Henry S. Parker, a faculty member of the Department of Biology at Southeastern Massachusetts University, has worked in the field of ocean science for more than fifteen years as a researcher, teacher, and U.S. Navy diver. He has published popular and scientific articles in the fields of aquaculture and algal biology and holds a Ph.D. in Biological Oceanography from the University of Rhode Island.

Exploring the)
OCEANS

An Introduction for the Traveler and Amateur Naturalist

HENRY S. PARKER

A SPECTRUM BOOK

Prentice-Hall, Inc., Englewood Cliffs, New Jersey 07632

Library of Congress Cataloging in Publication Data

Parker, Henry S.
    Exploring the oceans.

    (PHalarope books)
    "A Spectrum Book."
    Bibliography: p.
    Includes index.
    1. Oceanography—Popular works. 2. Marine
biology—Popular works. I. Title.
GC21.P34   1985   551.46   85-602
ISBN 0-13-297714-1
ISBN 0-13-297706-0 (pbk.)

*To Sue, James, and John
and many more
years of exploring together.*

1   2   3   4   5   6   7   8   9   10

Editorial/production supervision by Chris McMorrow
Page layout by Alice Mauro
Chapter opening design by Sue Maksuta
Cover design by Hal Siegel

This book is available at a special discount when ordered in
bulk quantities. Contact Prentice-Hall, Inc., General
Publishing Division, Special Sales, Englewood Cliffs, N.J. 07632.

Prentice-Hall International (UK) Limited, *London*
Prentice-Hall of Australia Pty. Limited, *Sydney*
Prentice-Hall Canada Inc., *Toronto*
Prentice-Hall Hispanoamericana, S.A., *Mexico*
Prentice-Hall of India Private Limited, *New Delhi*
Prentice-Hall of Japan, Inc., *Tokyo*
Prentice-Hall of Southeast Asia Pte. Ltd., *Singapore*
Whitehall Books Limited, *Wellington, New Zealand*
Editora Prentice-Hall do Brasil Ltda., *Rio de Janeiro*

ISBN 0-13-297714-1

ISBN 0-13-297706-0 {PBK.}

Contents

Foreword

OVER THE YEARS THE SUBJECT OF OCEANOGRAPHY AS A WHOLE has matured into a brilliant and fascinating field of study. To approach the topics that comprise oceanography is a demanding task and one that requires a broad scope of understanding and a sensitive pen. There is room in this science for the dissemination of knowledge at all levels. Certainly there is plenty of room for all to make knowledge of the oceans part of their daily lives.

Henry Parker has undertaken this task with new approaches and in a fine style. Oceanography is a discipline often labeled "man's last frontier." Henry's book has taken a very positive step toward leading us to this frontier and demystifying the oceans. It affords its readers new insights into this world of life, darkness, cold, and pressure.

As a teacher, I have always felt that the fascination of the world's oceans should be shared with all who have an interest. From the changing climates and their effects on water levels and sediments we gain insights into our planet's past. New discoveries come to life yearly. Startling concepts in chemistry have been made known through the study of hydrothermal vents; formerly unknown species have been catalogued at these sites. The study of plate tectonics gives us clues as to the future of the earth's continents and ocean basins. Through these studies we can begin to predict various geological events that will have an impact on our lives. The relationship of oceans, atmosphere, and climate gives us knowledge of past and future climates. The profound consequences for the climate and the ocean ecosystem were brought home dramatically in the 1982–1983 El Niño events that caused damage in the millions of dollars. The United Nations Law of the Sea Conferences struggled with ownership and uses of the world's oceans. As our studies move forward, we become more aware of the oceanic potential, and this book helps to bring us closer to the realization of the oceans' importance.

The future of the world's oceans depends on a well-educated citizenry. This is not to assume that all citizens must become scientists; rather, we all must continue to gain insights into the complex totality of the ocean ecosystems and how these great bodies of water affect our lives and we their well-being. In this volume there is valuable information both for those with an academic sense of purpose and for those who wish to better understand the oceans for their own sense of wonder.

Henry's book has much to offer its readers. It presents oceanography in a new light and a clear format. As you read it, remember that the knowledge you take from it can enrich your life and make you a valuable citizen in the world of the oceans.

PRENTICE K. STOUT
Executive Director
National Marine Education Association

Preface

THE OCEANS HAVE A FASCINATION FOR THOSE OF US WHO ARE fortunate enough to have been exposed to their mysteries. They do, after all, cover two-thirds of the earth's surface. Yet for most of us it is so easy to forget their vital role on our planet. Having had the good fortune to have helped awaken the interest of students exposed to oceanography for the first time, I find that they are invariably fascinated once they are introduced to the most fundamental concepts and descriptions in the ocean sciences.

You need not be a science student to understand and appreciate the concepts and descriptions set forth in this book. Indeed, you need not be a "student" at all. The book is designed to impart basic ocean science in a clear and readable fashion, without sacrificing the technical substance required for the serious student. It therefore is organized to serve as an introductory text for the college or high school student, and as a guidebook for the naturalist or traveler. To achieve this objective, I have chosen to use a narrative style and an exploratory approach wherever feasible. If, having read the book, you come away with a new perspective on the oceans and oceanography and feel as though your appetite for further exploration has only been whetted, then this book will have achieved its purpose.

ACKNOWLEDGMENTS

It is virtually impossible to acknowledge adequately all the many people who have had a hand in shaping this book, but I want to extend my grateful appreciation to several who have been particularly helpful and supportive.

First and foremost, I would like to express my gratitude to Nelson Marshall for his unfailing support and encouragement, not only in the preparation of this book but throughout my professional career.

I have imposed upon the knowledge and talents of several individuals in asking them to review and criticize sections of the manuscript. This they have done willingly and conscientiously, and the manuscript has benefited accordingly. The reviewers included Alan Bates of the Chemistry Department of Southeastern Massachusetts University (sections on chemistry) and Leah Curran, Eric Johnson, Sandy Moss, Frank O'Brien, Jefferson Turner, and Steve Warburton of the Biology Department of Southeastern Massachusetts University (sections on biological oceanography). Additional reviewers included Nelson Marshall of the Graduate School of Oceanography, University of Rhode Island (chapters on biological oceanography); Peter Meyer (chapters on geological oceanography) and Ray Schmitt (chapters on physical oceanography) of the Woods Hole Oceanographic Institution; and Phil LeBlanc and Wendell Hahm of the National Marine Fisheries Service, Northeast Fisheries Division, Woods Hole. Prentice Stout, of the University of Rhode Island and Executive Director of the National Marine Education Association, kindly consented to read and constructively criticize the entire manuscript. Although the suggestions of all these reviewers were extremely helpful and incorporated wherever feasible, mistakes may remain and are my own responsibility.

A number of individuals and institutions provided art work or photographs. I am particularly grateful to Bill Shattuck for his superb illustrations that grace the beginning of each chapter and to Susan Barrow Parker for her prodigious efforts and talent represented in the many outstanding line drawings that effectively illustrate the concepts described. Thanks also go to Paul Johnson, John Sieburth, and Paul Hargraves of the Graduate School of Oceanography, University of Rhode Island; to Jim Sears of Southeastern Massachusetts University; to John Huguenin of the Massachusetts Maritime Academy; to Sippican Ocean Systems, Inc.; to the Woods Hole Oceanographic Institution; and to the National Marine Fisheries Service for excellent photographs provided. Jim Feeley of the Audio/Visual Department of Southeastern Massachusetts University labored long and hard in producing the lettering in the illustrations, and Manny Pereira of the same department helpfully provided vital photographic reproduction assistance. Credits for illustrations reproduced from other sources are provided with the figure captions.

I am grateful to my father, Henry S. Parker, Jr., for his excellent editorial advice which greatly improved the clarity of portions of the manuscript.

Rita Sasseville and Stelle Mannion of Southeastern Massachusetts University and Andrea Lentz provided much-appreciated typing or clerical assistance. Ben Lentz provided support and help in many ways, as did staff at the Brewer Engineering Laboratories, Marion, Massachusetts.

The staff at Prentice-Hall has been a pleasure to work with, and I particularly appreciate the advice, support, patience, and assistance of the PHalarope Series Editor, Mary Kennan, and her assistants Stephanie Kiriakopoulos and Laura Likely, and the production editor, Chris McMorrow. It can't be easy working with a novice author, and I probably taxed their patience more than most.

Finally, and most important, I want to express my extraordinary gratitude to my wife, Susan, for everything she contributed to the book. Her line drawings speak for themselves, but without her unfailing support, patience, help in a myriad of ways, and love this book would never have been written.

Figures 2.6, 4.13, and 7.12 were adapted from Peter K. Weyl, *Oceanography: An Introduction to the Marine Environment,* © 1970, John Wiley and Sons, Inc., New York.

Figures 2.14, 4.1, 4.8, 4.11, 11.19, 11.24, 11.28 are from Alyn C. Duxbury and Alison Duxbury, *An Introduction to the World's Oceans,* © 1984, Addison Wesley Publishing Co., Reading, Massachusetts, and are reprinted with permission.

Figures 3.1, 7.1, 8.17, 8.18, 9.5, 9.6, 9.9, 9.22, 9.23, and 10.11 were adapted from Sverdrup, Johnson, and Fleming, *The Oceans: Their Physics, Chemistry, and General Biology,* © 1942, renewed 1970. Adapted or reprinted by permission of Prentice-Hall, Inc., Englewood Cliffs, N.J.

Figures 3.3, 3.4, 3.8, 4.6, and 8.10 were based on M. Grant Gross, *Oceanography: A View of the Earth,* 3rd Ed., © 1982. Reprinted by permission of Prentice-Hall, Inc., Englewood Cliffs, N.J.

Figures 4.12, 5.8, 6.1, 6.2, 7.4, 8.6, 8.8, 8.11, and 8.12 are used with the permission of Keith Stowe and John Wiley & Sons, Inc., Publishers. They are from Keith Stowe, *Ocean Science,* © 1983 (2nd Ed.) John Wiley & Sons, N.Y.

Figure 5.4a is adapted from J. E. Brady and G. E. Humiston, *General Chemistry: Principles and Structure,* 2nd Ed., © 1978 John Wiley & Sons, Inc., N.Y. and is used by permission.

Figures 5.9, 5.12, 7.16, 8.9, 8.21, 8.26, and 10.1 are adapted/reprinted from William A. Anikouchine and Richard W. Sternberg, *The World Ocean: An Introduction to Oceanography,* © 1973, and are used by permission of Prentice-Hall, Inc., Englewood Cliffs, N.J.

Figures 7.8 and 7.11 are adapted with permission from P. Tchernia, *Descriptive Regional Oceanography,* © 1980 Pergamon Press Ltd.

Figures 7.13, 7.14, 7.15, and 7.17 are adapted with permission from G. L. Pickard and W. J. Emery, *Descriptive Physical Oceanography,* 4th Ed., © 1982 Pergamon Press Ltd.

Figure 9.39 is adapted with permission from Parsons et al., *Biological Oceanographic Processes,* 2nd Ed., © 1977 Pergamon Press Ltd.

Figure 9.4 is adapted from James L. Sumich, *An Introduction to the Biology of Marine Life*, 3rd Ed., © 1976, 1980, 1984 Wm. C. Brown Publishers, Dubuque, Iowa.

Figure 9.40 is taken from R. V. Tait, *Elements of Marine Ecology*, 3rd Ed., © 1981 Butterworths Scientific Ltd., Sevenoaks, Kent, England.

Figures 10.8a, b, and c, and 10-12 are reprinted with the permission of the Carlsberg Foundation, Copenhagen, Denmark.

Chapter One

Introduction

LIFE ON EARTH OWES ITS EXISTENCE TO THE PRESENCE OF WATER on the surface of the planet. We take this water for granted, but the liquid envelope surrounding our globe is unique—no other planets in our solar system are so blessed.

The presence of water on earth is remarkable in itself, but its abundance is even more noteworthy. Seventy-one percent of the globe's surface is covered with water, with virtually all this water being in the oceans. Indeed, the suggestion that our world would more properly be called "Planet Ocean" certainly makes sense.

As extensive and important as the ocean environment is, we know surprisingly little about it. Vast, dark, and deep, it is difficult to investigate and to interpret. The oceans yield their secrets grudgingly.

The difficulties have not discouraged scientists from focusing their research on the oceans. Today, the discipline of *oceanography* is a global endeavor occupying the attention of legions of investigators from a wide variety of fields.

THE SCIENCE OF OCEANOGRAPHY

Oceanography may be defined simply as the study of oceanic processes, but the simplicity of that definition is misleading. The complexity of the ocean environment and the closely coupled relationship of the oceans to the atmospheric and terrestrial environments demand that the study of the oceans be approached from a multidisciplinary perspective. Indeed, among those disciplines represented by oceanographers today may be found virtually every branch of science currently practiced.

Traditionally, four broad areas of oceanography are recognized: *geological oceanography*, the study of the structure and processes of the sea floor; *chemical oceanography*, the investigations of the chemical compositions and reactions in the oceans; *physical oceanography*, which is concerned with the nature and properties of moving water; and *biological oceanography*, whose focus is on the living organisms of the seas. However, there is considerable overlap between these areas: the discipline of marine biogeochemistry is a case in point.

We all have our own perceptions of why it is important to study the seas, but some reasons particularly stand out. First and foremost, the ocean environment is vast. Life evidently evolved in the oceans, and the first nine-tenths of the history of life occurred in them. There is 300 times as much habitable space in the oceans as in the terrestrial environment. The oceans supply many of our food, chemical, and material resources and promise to supply much more in the future. They are both historically and presently a crucial medium for transportation and communication and are militarily and strategically vital. They are an increasingly important source of energy. The dynamics of the oceans are integrally linked with those of the atmosphere and, as such, govern the earth's long-term climate and short-term weather. Our growing knowledge of the processes of the sea floor has enabled us to understand larger, global geological processes. Finally, it is reason enough to study the oceans simply because they are fascinating.

The History of Oceanography

Oceanography is a young science, but its roots are very old. The earliest students of the ocean environment were the ancient mariners, whose interests were primarily navigational. Long before modern oceanography, the accumulated knowledge of ocean surface currents and wind patterns gleaned from thousands of maritime voyages was surprisingly sophisticated and had led to the development of effective navigational instruments. The tradition of studying the oceans from a navigational and geographic perspective was carried well into the eighteenth century and exemplified by the voyages of the Portugese and Spanish seamen in the fifteenth and sixteenth centuries and by the three exploratory investigations of the Pacific Ocean carried out by Captain James Cook between 1768 and 1780.

By the seventeenth century the study of the oceans had begun to be truly scientific, with its objective the search for knowledge for its own sake. In 1665, Britain's Royal Society, drawing on the work of Robert Boyle, Sir Isaac Newton, and Robert Hooke, published detailed instructions for investigations of ocean salinity, temperature, depth, tides, and currents. In the seventeenth and eighteenth centuries, major contributions to ocean science were made by several chemists and physicists. In 1776, Antoine Lavoisier, a French chemist, published a detailed analysis of sea-water chemistry. In 1820, the Swiss-born London physician, Alexander Marcet, correctly observed that the proportions of salinity constituents of sea water were essentially the same everywhere in the oceans. In 1835,

Gaspard Gustave de Coriolis was the first scientist to elucidate the effects of the earth's rotation on moving fluids. Matthew Fontaine Maury, superintendent of the U.S. government's Depot of Charts and Instruments in the mid-nineteenth century, provided a systematic synthesis of ocean currents and weather conditions in his *Physical Geography of the Sea*, published in 1855. Interest in marine biology was also flourishing by the eighteenth century, as exemplified by the popularity of Reverend William Spiers's *Rambles and Reveries of a Naturalist*, which was based on his collecting trips to the sea shore.

Though often meritorious, early scientific investigations of the sea were disjointed, unrelated efforts which lacked the interdisciplinary approach characterizing oceanography today. However, by the mid-nineteenth century enough discrete bits of information were available to suggest that the marine environment was a complex system characterized by closely coupled relationships between its components. In part because of this evolving recognition, and also because of growing interest in the deep-sea environment and concern about erratic fish catches, incentives existed for scientific expeditions whose sole purpose was to learn more about the oceans. The globe-encircling, British-sponsored voyage of the H.M.S. *Challenger*, from 1872 to 1876, was the first true oceanographic expedition and essentially marked the birth of oceanography as an interdisciplinary science. The *Challenger* expedition is described in detail in Chapter 10.

Following the *Challenger* expedition a number of scientific voyages were launched. The late 1800s saw several investigations of physical oceanography by Scandinavian physicists and chemists. Notable among these was the 1893 voyage of the Norwegian explorer, Fridtjof Nansen, to study the circulation of the Arctic Ocean. His ship, the *Fram*, was specially constructed with a reinforced hull capable of withstanding being crushed by ice. In addition to confirming the absence of a northern continent, Nansen fortuitously discovered, by keeping careful records of the position of the *Fram* after it was frozen into floating ice, that the ocean currents flowed to the right of the wind direction. The Swedish physicist, V. W. Ekman, subsequently confirmed that this was due to the effect, as seen in earlier observations by de Coriolis, of the earth's rotation on moving fluids, including currents (see Chapter 6).

Ekman collaborated with his compatriot, the chemist Otto Pettersson, in a series of systematic hydrographic studies of Scandina-

FIGURE 1.1. The Woods Hole Oceanographic Institution. (Courtesy of Woods Hole Oceanographic Institution. Photo by Vicky Cullen.)

vian water masses to test the hypothesis that variable herring abundance was related to water temperature and salinity. This work helped launch the International Council for the Exploration of the Sea (ICES) in 1902. Supported by the Swedish king, the Council promoted international, interdisciplinary studies of oceanographic conditions thought to affect the North Atlantic fisheries.

From 1925 to 1927, the German research vessel *Meteor* conducted a major investigation of the physical, chemical, biological, and geological properties of the Atlantic Ocean. Using state-of-the-art equipment for that time, including newly developed echo sounders, *Meteor* sampled the sea floor and worked up vertical profiles of water temperature, salinity, and dissolved oxygen along a series of transects across the ocean. Her voyage is said to have ushered in the modern era of oceanography, since it was the first to undertake a systematic study of an entire ocean.

The United States's activities in oceanography accelerated with the establishment of several outstanding research and educational institutions in the first half of the twentieth century. The Scripps Institution of Oceanography (known as "Scripps") of the University of California at San Diego was founded in 1905 and has become an international leader in marine sciences since then. The equally distinguished Woods Hole Oceanographic Institution (WHOI) was established in 1930 on the south shore of Cape Cod, Massachusetts (Figure 1.1). As in many oceanographic institutions, much of WHOI's early work was carried out on sailing vessels, notably the R/V *Atlantis* (Figure 1.2). In 1948, the Lamont–Doherty Geological Observatory (now part of Columbia University) was founded. A number of United States universities, including the University of Washington, Oregon State University, the University of Rhode Island, Texas A. and M. University, the University of Texas, the University of Hawaii, the University of Miami, and Florida State University, now have major programs in oceanography with seagoing research vessels (see Chapter 2). Most U.S. oceanographic work is supported by the federal government, principally through the National Science Foundation (NSF), the U.S. Navy, the U.S. Coast Guard, the National Oceanic and Atmospheric Administration (NOAA), and the U.S. Geological Survey (USGS). These agencies also own and operate a number of seagoing research vessels.

Oceanography Today and Tomorrow

Oceanographic research today is an international endeavor characterized by joint programs between institutions and nations. Cooperative international studies began in earnest in the 1950s with the

FIGURE 1.2. R/V *Atlantis* underway. (Courtesy of Woods Hole Oceanographic Institution. Photo by Gerry Metcalf.)

International Geophysical Year (IGY) in 1957–1958 and the International Indian Ocean Expedition, overseen by the United Nations Educational, Scientific, and Cultural Organization (UNESCO) from 1959 to 1965. The most notable achievement of the IGY was the detailed mapping of the ocean basins conducted by scientists from Lamont–Doherty. Their work contributed significantly to developing ideas regarding plate tectonics (see Chapter 4).

The Deep Sea Drilling Project (DSDP) was initiated in 1968 (funded by NSF and managed by Scripps) to test the hypothesis of sea-floor spreading and plate tectonics, mechanisms proposed to account for continental drift. In 1975, the project drew in the support and participation of the United Kingdom, France, Japan, the Soviet Union, and the Federal Republic of Germany and became the International Program of Ocean Drilling. The drilling platform has been the *Glomar Challenger*, a specially constructed vessel with dynamic positioning capability and modified oil-drilling equipment which allow it to recover cores of solid rock from the deep sea floor. *Glomar Challenger* was recently retired, but its place has been taken by a newer and larger vessel.

In 1968, U.S. President Lyndon Johnson endorsed and the United Nations supported the launching of the International Decade of Ocean Exploration (IDOE) for the 1970s. The purpose of IDOE, as expressed by the International Oceanographic Commission (IOC) of UNESCO was "to increase knowledge of the ocean, its contents and the contents of its subsoil, and its interfaces with the land, the atmosphere, and the ocean floor and to improve the understanding of processes operating in or affecting the marine environment, with the goal of enhanced utilization of the ocean and its resources for the benefit of mankind."

The cooperative research projects which resulted from the IDOE involved thirty-six nations. Notable studies included Project FAMOUS (French-American Mid-Ocean Undersea Study), which employed manned submersibles to investigate geological processes responsible for the formation of ocean basins; studies of the Nazca Plate in the eastern Pacific conducted by scientists from the United States, Bolivia, Canada, Chile, Colombia, Ecuador, Peru, and the USSR; studies of the Galapagos Rift and East Pacific Rise which resulted in the discovery of an unusual and highly productive community of organisms associated with submarine hot springs 4 kilometers below the sea surface; investigations of manganese nodules, which are commercially valuable, fist-sized, polymetallic lumps of

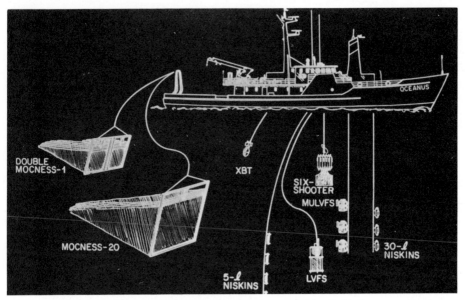

FIGURE 1.3. Representative sampling devices used by R/V *Oceanus*. Devices would not all be deployed simultaneously. (By Peter Wiebe. Courtesy of Woods Hole Oceanographic Institution.)

minerals which accumulate in some areas of the sea floor; the Coastal Upwelling Ecosystem Analysis (CUEA), which involved studies of the factors affecting the rising up of deep, nutrient-rich waters which are responsible for the high productivity of some coastal fisheries; several investigations of the effects of the oceans on global climate; and a number of studies of ocean pollution. The concept of the IDOE has been continued into the 1980s as the Coordinated Ocean Research and Exploration Section (CORES), supported by NSF.

FIGURE 1.4 Research submersible *Nekton* operated by the National Marine Fisheries Service. (Photo by James Sears.)

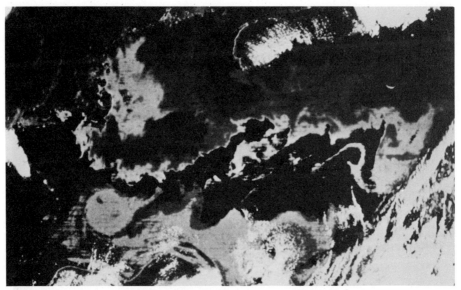

FIGURE 1.5. Infrared satellite photograph of the North Atlantic Ocean. Gulf Stream ring and meanders are visible in lower center as grayish areas. (Courtesy of Woods Hole Oceanographic Institution.)

Oceanography today is still largely ship-based, but this is changing. Sampling devices have become increasingly efficient, largely through the use of electronic equipment (Figure 1.3). Several ships often work together so that processes occurring over a large area of the ocean can be studied at one time. Oceanographic horizons have been vastly extended through the use of manned and unmanned research submersibles (Figure 1.4), underwater camera systems, and remote sampling and sensing techniques, including the use of satellites (Figure 1.5).

Major future developments in oceanography are likely to come through increased reliance on remote sensing techniques. It has been recognized for some time that subtle color variations in the oceans' surface waters, discernible from even a low-flying aircraft, may tell us much about the physical, chemical, and biological properties of water masses. For example, through aerial surveillance in 1956 Sir Alister Hardy was able to distinguish the boundary between the English Channel and the Irish Sea by recognizing the differences in color due to abundant salp (small, transparent plankton which are primitive relatives of vertebrates) populations in the latter water body. The use of infrared radiometers to detect temperature differences in air and water has been in effect since World War II and was applied to oceanographic investigations as early as 1955, when WHOI scientists used a radiometer to track the Gulf Stream.

It was only natural to apply these observations to the use of satellite platforms, which can accurately scan large areas of the sea surface in short periods of time. The first use of satellites to observe oceanic phenomena occurred during the Mercury space program in the early 1960s. Today, satellites employ several remote-sensing devices to measure ocean temperatures, primary productivity, winds, waves, and slight differences in sea-surface height. Satellites also receive and transmit (to shore facilities) data from the oceanographic buoys and unmanned platforms. More detailed discussion of the use of satellites in oceanography can be found in Chapter 6.

Oceanography is a complex science which will play an increasingly important role in our evolving understanding of the world in which we live. Come on board for an introductory exploration of this vast and fascinating environment.

SUGGESTED READINGS

The History of Oceanography

BAILEY, H. S. 1977. The voyage of the *Challenger*. In *Ocean science*. San Francisco: Freeman.

DEACON, M. 1971. *Scientists and the sea, 1650–1900: a study of marine science*. London and New York: Academic Press.

SCHLEE, S. 1973. *The edge of an unfamiliar world: a history of oceanography*. New York: Dutton.

Research Vessels

NELSON, S. B. 1971. *Oceanographic ships, fore and aft*. Washington, D.C.: U.S. Government Printing Office.

RYAN, P. R., ed. 1982. Research vessels. *Oceanus* 25(1). (A collection of articles.)

Satellite Oceanography

MACLEISH, W. H., ed. 1981. Oceanography from space. *Oceanus* 24(3). (A collection of articles.)

General References

ANIKOUCHINE, W. A., AND STERNBERG, R. W. 1981. *The world ocean: an introduction to oceanography*. 2nd ed. Englewood Cliffs, N.J.: Prentice-Hall.

DRAKE, C. L., IMBRIE, J., KNAUSS, J. A., AND TUREKIAN, K. H. 1978. *Oceanography*. New York: Holt, Rinehart and Winston.

DUXBURY, A. C., AND DUXBURY, A. 1984. *An introduction to the world's oceans*. Reading, Mass.: Addison-Wesley.

GROSS, M. G. 1982. *Oceanography: a view of the earth*. 3rd ed. Englewood Cliffs, N.J.: Prentice-Hall.

MCCORMICK, J. M., AND THIRUVATHUKAL, J. V. 1981. *Elements of oceanography*. 2nd ed. New York: Saunders.

MEADOWS, P. S., AND CAMPBELL, J. L. 1978. *An introduction to marine science*. New York: Wiley.

1977. *The Rand McNally atlas of the oceans*. New York: Rand McNally.

1971. *Oceanography*. San Francisco: Freeman. (Readings from *Scientific American*.)

1977. *Ocean science*. San Francisco: Freeman. (Readings from *Scientific American*.)

STOWE, K. 1983. *Ocean science*. 2nd ed. New York: Wiley.

SVERDRUP, H. U., JOHNSON, M. W., AND FLEMING, R. H. 1942. *The oceans: their physics, chemistry, and general biology*. Englewood Cliffs, N.J.: Prentice-Hall.

THURMAN, H. V. 1981. *Introductory oceanography*. 3rd ed. Columbus, Ohio: Merrill.

TUREKIAN, K. H. 1968. *Oceans*. Englewood Cliffs, N.J.: Prentice-Hall.

WEISS, H. M., AND DORSEY, M. W. 1979. *Investigating the marine environment: a sourcebook*. 3 volumes. Groton, Conn.: Project Oceanology, Avery Point.

WEYL, P. K. 1970. *Oceanography: an introduction to the marine environment*. New York: Wiley.

Chapter Two

A Day Aboard a Research Vessel

HAVE YOU EVER BEEN TO SEA? THOSE WHO HAVE KNOW THAT there is a rhythmic beauty about a quiet night on the ocean. On a darkened ship's bridge there is a womblike security, accentuated by the pulsing throb of the power plant far below. Seas glide by in monotonous procession, and uninterrupted darkness stretches into eternity. On such evenings, a scientist might momentarily forget that there is work to do.

It is a sobering discovery for initiate oceanographers that it is seldom calm at sea and that much of the work is done at night. Difficult, demanding, and tedious work is required before one can hope for significant results. Of all environments on earth, the oceans, because they are vast, dark, and turbulent, are probably the most difficult to study. Consider the problem of obtaining representative samples of marine life in the deep sea. By one analogy it is like trying to sample life in the terrestrial environment by using a butterfly net towed by an airplane.

Nonetheless, during night or day, in calm waters or a maelstrom, the seas must be sampled with efficiency and dispatch. Ship-based oceanography is simply too expensive to permit waste of time.

The probing and sampling of the ocean depths is a relatively recent development. Prior to a century or so ago, humans had only the vaguest notion of the physical and biological processes occurring

FIGURE 2.1. R/V *Albatross IV.* (Courtesy of the National Marine Fisheries Service.)

in the oceans. Such fragmented knowledge as did exist was discovered largely accidentally—and often misinterpreted. However, the British *H.M.S. Challenger* expedition, begun in 1872, ushered in a new era of systematic and rigorous examination of the oceans and, as such, gave birth to the scientific discipline of oceanography.

Today, a small navy of over 700 seagoing research vessels cruises the world's oceans. United States, Japanese, and Russian ships account for 40 percent of the total. Of 113 research vessels operated by the United States, 43 are under federal or state jurisdiction, 45 are commercial, and 25 are operated by universities. They range in size from 24 to 92 meters long and carry crews of 5 to 25 individuals. The largest can stay at sea for a month or more at a time and can accommodate over two dozen working scientists, whose work encompasses most of the major scientific disciplines. On long cruises, regular replacements of scientists may occur.

AN OCEANOGRAPHIC CRUISE

What are research vessels like, and what kind of work goes on aboard them? You might imagine yourself as the chief scientist on a winter oceanographic cruise in the North Atlantic. Your vessel is the R/V *Albatross IV* (Figure 2.1), a federal government ship operated by the National Marine Fisheries Service (NMFS) of the National Oceanic and Atmospheric Administration (NOAA) out of Woods Hole, Massachusetts. She is of intermediate size for a research vessel, with a length of 57 meters, a beam (the width of a ship at its widest part) of 10 meters, a draft (the depth of water required to float) of 5 meters, and a displacement (the weight of water displaced) of 1089 tons. She is designed to carry 7 licensed officers, 15 crew members, and 15 scientists and can operate for up to two weeks without refueling, at an average cruising speed of 12 knots (13.8 miles per hour). A single cruise may last six months with periodic port visits.

Among the oldest of U.S. research vessels, *Albatross IV* was launched in 1962, but her diesel power plant, her propulsion, and her deck machinery are relatively modern, and her electronic and scientific equipment are state-of-the-art. Rather squat and beamy, her upper decks bristling with booms, winches, and antennae, she lacks the sleek grace of a destroyer or the classic elegance of the century's earlier, yachtlike sailing research vessels (Figure 2.2). She is a

FIGURE 2.2. R/V *Atlantis*.
(Courtesy of Woods Hole
Oceanographic Institution.)

workhorse, purely and simply, with only her white hull, a trademark of government research vessels, softening her otherwise harshly utilitarian features.

A Study of the New York Bight

Albatross IV's normal arena is the Atlantic coast of the United States, where her crew conducts biological and fisheries research. In recent years much of her work has focused on the New York Bight, a 50,000 square kilometer area of the continental shelf south of Long Island and east of New Jersey (Figures 2.3 and 2.4). For nearly a century, the Bight has been a dumping ground for the New York–New Jersey metropolitan region, currently inhabited by 20 million people. Every year about 9 million tons of particulate, liquid, and solid wastes pour into its waters from a variety of sources, including the Hudson River system, sewer outfalls, runoff from land, and atmospheric fallout. Much of the waste is towed out on barges. The Bight receives about 65 percent of all industrial wastes, 85 percent of all municipal sludges, and 90 percent of all acid wastes dumped in U.S. coastal waters by vessels. It has been estimated that the amount of sewage sludge alone dumped in one year would cover the entire surface of New York City's Central Park to a depth of 1.2 meters.

What is the biological impact of this enormous volume of waste in a relatively confined area? Only within the past 10 to 15 years have we begun to seriously investigate it. Much of the research has been carried out by *Albatross IV*. The purpose of this cruise, one of

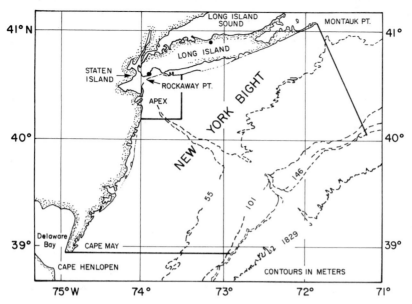

FIGURE 2.3. The New York Bight. (Redrawn from N.O.A.A. Technical Memorandum OMPA-6. MAY 1981. Fig. 1.)

FIGURE 2.4. Dump sites in the New York Bight Apex. (From N.O.A.A. Professional Paper 11, December 1979.)

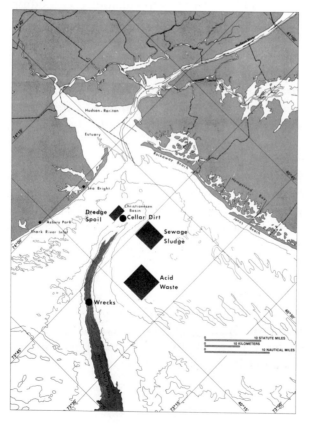

a series in the OCEAN-PULSE pollution studies program, is to gather and analyze biological, chemical, physical, and geological data which will contribute to our understanding of the impact of this unprecedented waste disposal.

As chief scientist of this cruise you have the job of overseeing studies of the water column and sediments in a section of the New York Bight allocated for sewage-sludge disposal. The section is about 25 kilometers square and located about 20 kilometers south of Long Island and east of New Jersey. It is one of five separate sites designated for the disposal of specific wastes. The others receive dredged mud, rubble and debris, wrecks, and acid wastes.

Cruise Preparations

The cruise will be a short one—only one week—but months have gone into its planning and preparation. Most important, scientists have analyzed data from previous dump site cruises. Based on these analyses plus evaluation of other factors—the availability of appropriate scientists, limitations of ship capability and equipment, and seasonal/weather considerations—two specific objectives have been formulated for this cruise. First, the oceanographers will examine midwinter physical, chemical, biological, and geological characteristics of the dump site and compare them with those of a "clean" control site 50 kilometers distant. Second, the scientists will monitor the site's short-term responses to the dumping of a fresh load of sewage sludge, again comparing these responses to the simultaneous "behavior" of the control site. Because two sites will be monitored at the same time, two research vessels are needed. The other will be the University of Rhode Island's *Endeavor*, slightly smaller but about 15 years newer than *Albatross IV*.

The scientists have prepared for the cruise by assembling, installing, and testing scientific equipment, ordering required supplies and chemicals, and working out the precise details of the research program during a number of precruise meetings. The ship's crew has been equally busy. They have secured fuel, fresh water, food, and stores. They have also conducted required repairs, maintenance, and operating checks on the ship's power plant, deck machinery, and electronics gear. You personally ensured that all required scientific gear was brought aboard and properly secured prior to getting underway. Sailing orders were issued, and the ship's

Executive Officer briefed you and the other scientists on rules and regulations while at sea. You in turn briefed all scientists on cruise objectives and procedures and verified that the ship's captain has a clear understanding of the proposed scientific program.

Effective communications with the captain are especially important. The chief scientist is responsible for the scientific conduct of the cruise, but the captain is responsible for the overall safety of the vessel and her occupants. A good chief scientist will respect and accept the captain's ultimate responsibility, just as a good captain will not interfere with a scientific program unless the vessel appears endangered.

Underway

Let's pick up the cruise on February 5, the fourth day out of port. Departure from Woods Hole occurred without fanfare at dawn on February 2. There have been no problems so far, and the initial dump site monitoring program has progressed without incident. At 0530 (nautical time runs on a 24-hour clock), a prearranged knock on your stateroom door jolts you out of your fitful sleep. In contrast to the Spartan and cramped quarters of other scientists and crew, staterooms of chief scientists are quite comfortable and spacious. Ironically, their occupants seldom have time to appreciate them. Today is no exception.

Groggily shaking the sleep from your eyes, you lurch to the porthole. The motion of the ship has become livelier since you turned in, and a glance outside tells why. Wind and seas have picked up considerably overnight. Dressing quickly, you make your way to the ship's mess (dining room), down a platter of hot cakes and sausages, and assemble the scientists for a briefing on the day's activities. Three bargeloads of sewage sludge will be disgorged into the center of the dump site at 0800, and an intensive sampling program will begin thereafter.

The Research Program

For the next three days, the scientists will monitor a number of variables at three different stations in the dump site in order to determine the distribution patterns of the discharged sewage and the effects of the discharge on light attenuation, water chemistry, and the nature and chemical composition of bottom sediments. Coupled with observations on the abundance and composition of marine organisms in the area, this information will contribute to understanding of the impact of the sewage dumping on marine life and physical and chemical properties of the New York Bight area.

Specific observations at each station will include the following: Physical oceanographers will determine atmospheric conditions and sea state. They will also monitor surface and subsurface salinity, temperature, light intensity, and current velocity and direction. Chemical oceanographers will take water samples at a series of depths. Each sample will be analyzed for dissolved oxygen content (likely to be diminished by decomposing organic matter in sewage), concentrations of nutrients and heavy metals characteristically associated with sewage sludge, and pH. Biological oceanographers will retrieve water samples with special sampling bottles, and will tow fine- and coarse-meshed nets through the waters at various depths to determine the composition and abundance of *plankton* (floating marine life) and *nekton* (swimming marine life). They will also dredge the sea floor to make similar assessments of the *benthos* (bottom life). Geological oceanographers will determine bottom topography and take vertical cores of the sediments on the sea floor to determine their composition, grain size, porosity, and chemical properties. All measurements and analyses will be compared with values obtained before dumping and with values at the control site, where the same work is being carried out by *Endeavor*.

Having reviewed the basic sampling program with the other scientists, you carefully establish the sequence of events for activities at the first station, stressing the importance of timing. After verification of the precise location, the physical and chemical oceanographers will undertake their measurements first, while the ship remains fixed in position 500 meters downcurrent of the center of the dump site. The ship will then steam slowly along predetermined transect lines while the biologists conduct tows for plankton, nekton, and benthos. The geologists will take a final series of cores on the last day of the cruise in order to compare sediment characteristics with those represented by cores taken on the first day at the station, prior to sludge dumping. These procedures will be repeated for each station. R/V *Endeavor's* results from the "clean" area will be compared with those from the area affected by sewage.

The briefing complete, the scientists disperse to one of the ship's five separate laboratories or to the weather decks to make final preparations. Cursing the worsening sea conditions, you lunge up the ladder leading to the ship's bridge (the control center of the vessel). After requesting permission of the watch officer, you enter the glassed-in area. In the dim early morning light you make out the helmsman and the captain (all individuals described in this narrative are fictitious and not intended to represent any person, living or dead), who acknowledge your presence with a nod. A soft-spoken merchant mariner in his mid-50s, the captain is a graduate of the Massachusetts Maritime Academy and received his Master's license for seagoing vessels in 1962. After a brief stint aboard oil tankers ("like sailing a shoe box"), he has captained research vessels since the mid-1960s and has been master of *Albatross IV* for six years. This is your third cruise with him, and you like him and respect him.

Relations between ships' masters and chief scientists, while not always openly friendly, are usually based on mutual respect. Occasionally problems arise. More than once, tensions between the chief scientist and master have escalated to the level where the master has effected the removal of the chief scientist from the ship. Such personality clashes are inevitably to the detriment of productive research.

Peering out the bridge windows, you make out a tug towing three barges about 10 kilometers to the northwest. The dumping should take place on schedule. The crackling of the bridge UHF radio transceiver signals the initiation of voice communications between the tug and *Albatross IV*. It's time for you to get to work.

You make your way back to the chartroom just aft of the bridge and, with the aid of the LORAN (Long Range Navigation) and satellite navigation systems, verify the ship's position 1000 meters southwest of the first sampling station. Electronic navigational aids have largely obviated the need for celestial navigation, but the latter is still used when equipment breaks down or when there is substantial interference. The position plotted, you carefully back down the ladder to check the preparations of the other scientists. All is in order except for the stomachs of the two biologists whose rigging of a plankton net is frequently interrupted by dashes to the leeward (downwind) rail. Seasickness is a fact of life on any ship, but only if it is unusually severe can it be allowed to curtail required work. In fact, susceptible mariners routinely take advantage of several highly effective antiseasickness drugs developed in recent years.

Of more concern is the worsening weather. You make a quick trip back to the bridge to confer with the captain. The barometer is down, wind and seas have increased since 0600, and a light sleet is rattling against the bridge windows. The forecast is uncertain. A northeaster is gaining strength off Cape Hatteras, but there are some indications that it will move far enough east to only brush the New York Bight. If the weather deteriorates significantly, the ship will have to cease scientific operations, heave to (stop forward motion of the vessel by heading into the wind), and ride out the storm. You cross your fingers and decide, with the captain, to proceed with the sampling program.

FIGURE 2.5. CTD/rosette water sampler. See text for description of operation. (Courtesy of Woods Hole Oceanographic Institution. Photograph by Peter Wiebe.)

A nauseating stench assails the bridge. "We're downwind again," growls the captain. You glance at your watch—at least the barges were emptied on schedule. On deck, the starboard (right) side amidships is a flurry of activity. A large circular steel frame is being guided out of an open weather-deck door, propelled in an overhead track. Rubberized cable snakes from its top to a spool mounted on the bulkhead (wall), and a half dozen large, vertically aligned plastic water bottles rim its periphery. Each bottle is open at both ends, but large rubber stoppers are drawn back in a cocked position adjacent to the openings. This unlikely looking device is a *CTD* (Conductivity-Temperature-Depth profiler) with an attached water-sampling "rosette" (Figure 2.5). (Measurements of conductivity enable precise determinations of salinity, as described in Chapter 5.)

Standard equipment on most research vessels, the CTD (or its cousin, the *STD*, a Salinity-Temperature-Depth profiler) is one of the most important instrumental arrays currently available to sea-going oceanographers. As it is lowered into the depths, it can provide a continuous electronic record of the variables for which it is named as well as capture intact samples of water from a predetermined sequence of depths without risking contamination from waters above and below. The latter is accomplished by remotely triggering the closure of the caps of a particular bottle at a particular depth.

A *reversing thermometer* is attached to each bottle. With its mercury protected by a thick glass shield to eliminate the effects of pressure, it provides a precise temperature reading (to 0.01° C) at the depth at which the bottles actually close. The principle is elegantly simple. When the stoppers are triggered shut on a specific bottle, its thermometer is simultaneously inverted, breaking the column of mercury at a constricted loop in the stem. The mercury extruded above the loop at the time of inversion accumulates in the other end. The amount which accumulates corresponds to the temperature at the time of reversal. A graduated scale in the inverted end provides a calibrated reading. A reversing thermometer and its principle of operation are shown in Figure 2.6.

While the CTD is being moved into place, a scientist gets in position on a small platform projecting over the water. Appropriately called the *hero platform*, its function is to help ensure that

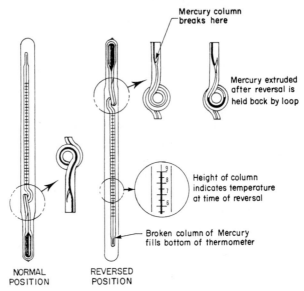

FIGURE 2.6. Reversing thermometer and principle of operation. (From Weyl, P. 1970. *Oceanography.* N.Y.: Wiley.)

The following labels appear on the figure:

Mercury column breaks here

Mercury extruded after reversal is held back by loop

Height of column indicates temperature at time of reversal

Broken column of Mercury fills bottom of thermometer

NORMAL POSITION

REVERSED POSITION

sampling instruments are lowered into the water well clear of the ship (Figure 2.7).

On the deck above the hero platform a massive spool of steel cable (the *hydrowire*) is being unwound from a *hydrographic winch* and reeved (led) through a block (pulley) attached to the end of a nearby boom. Hydrographic winches can handle over 9000 meters of cable and can lower and retrieve instruments weighing up to one-half ton. The end of the cable is shackled into a pad eye (securing ring) on the CTD. A signal is given, and the cable is slacked off carefully, allowing the instrument to be lowered steadily to an ultimate depth of 25 meters, just above the bottom, which averages 27 meters below the surface in this region. The *hydrocast* is underway.

In the meantime, the captain has positioned the ship 500 meters southwest of the actual dumping location. Maintenance of precise position, despite sea state, is crucial to the success of the measurements and the proper handling of the sampling gear. Fortunately, the captain is a fine seaman, and *Albatross IV* has controllable pitch propellers and auxiliary propulsion gear in the form of a through-hull *bow thruster* at the forward end of the ship. Normally retracted inside the hull when the ship is steaming, the bow thruster can be pivoted 360 degrees to help maintain the ship's position and minimize adverse effects of wind and waves. Despite such precautions, cables and winches sometimes fail because of overloading, improper maintenance, or deployment in excessively severe weather. The costs of such failures are high—up to $100,000 to replace a lost instrument array and $66,000 to purchase a new spool of hydrographic cable. Larger winches cost a half million dollars, and precision echo sounders up to $750,000.

Lost or damaged gear is but one of the factors contributing to the singularly high cost of oceanographic research. Based on a 1983 analysis, annual operating costs are nearly a million dollars for small vessels and up to 3 million dollars for large ships. In terms of daily operating costs for at-sea time, this translates into $3,700 to $10,500. Crew salaries (36 percent of the total) and fuel (21 percent) represent the largest single operating costs, but the costs of supplies and stores, overhead, shore-based marine support staff, food, overhaul,

FIGURE 2.7. Hero platform. (Courtesy of Woods Hole Oceanographic Institution.)

and maintenance also contribute significantly. These figures do not even include the salaries of scientists and technicians. All in all, ship operating costs represent 20 to 30 percent of the federal funds allocated for oceanographic research. Unfortunately, these costs are increasing at a faster rate than federal support. An inevitable consequence has been that three university ships had to be laid up in 1981–1982.

As the CTD disappears beneath the waves, you make your way to the electronics lab, one deck below the bridge. Every inch a sophisticated modern laboratory, its array of whirring computer discs and flashing instrument panels seems a far cry from the heavy work being done on deck. In here are located submarine acoustics equipment and a computerized data-acquisition-and-processing system which integrates navigational and oceanographic measurements. The control and display instruments for the CTD and other electronic monitoring and recording devices are also housed in the lab. The CTD operator is examining a growing strip of chart paper which records a continuous profile of temperature, salinity, and depth. You confirm with the operator that the intrument is programmed to trigger the closure of sampling bottles at 2, 5, 10, 15, 20, and 25 meters. At a lowering and retrieval time of approximately 2 meters per minute, the entire hydrocast will take about a half hour.

Launching an XBT and a Monitoring Buoy

While the CTD is being lowered, two other operations are occurring simultaneously. At the after end of the fantail (aftermost open weather deck), a scientist is launching an expendable device called an *XBT* (expendable bathythermograph), which can provide a profile of the temperature structure of the upper layers of the ocean. Shaped like a miniature torpedo and mechanically launched from a gunlike barrel (Figure 2.8), this nonself-propelled device sinks at a known rate. Behind it trails a thin copper wire which transmits temperature back to an electronic recorder on the ship and breaks when the instrument reaches the bottom or its depth limit. By knowing the sinking rate and time, the reader can relate the transmitted temperature to depth. Although it provides less accurate temperature data than the CTD, the XBT has the advantage that it can be used while the ship is underway. One XBT is launched every hour, around the clock, for the duration of the cruise.

(a)

FIGURE 2.8. Expendable bathythermograph (XBT). (a) Launching. (b) Underway (the device is not self-propelled). (Courtesy of Sippican Ocean Systems, Inc., Marion, Massachusetts.)

(b)

FIGURE 2.9. Launching of a meteorological monitoring buoy. (Courtesy of Woods Hole Oceanographic Institution. Photograph by Mel Briscoe.)

Further forward on the fantail, four oceanographers are preparing to launch a large multipurpose buoy from a davit on the starboard side. Once in the water, the buoy is moored in place by a long wire attached to an anchor on the sea floor. The buoy is designed to record data internally and to simultaneously telemeter it to a ship or satellite for relay to shore. Atmospheric sensors will measure wind speed and direction, temperature, and barometric pressure while underwater sensors monitor current speed and direction at five different depths as well as wave height and surface-water temperature. These data will help scientists interpret the dispersion patterns of the sewage discharged into the study area. The current buoy can remain in place indefinitely, with periodic maintenance and operating checks, and will be left on site at the completion of the cruise. A buoy similar to the one described is shown in Figure 2.9.

Light Measurements

Once the buoy has been successfully deployed, two scientists bring out a small instrument from a sturdy wooden box on the deck. With its calibrated scale and needle the device resembles a battery tester but is, in fact, a light meter called a *quantum/radiometer/photo-*

meter. Its operation is surprisingly simple. A probe on the end of a long rubber cable is lowered into the sea. The probe reads light intensity in photosynthetically active wavelengths (spectra that green plants can use for photosynthesis) and transmits it to the calibrated scale on deck.

As one scientist lowers the cable, the other records readings. At 10 meters the reading is zero. While the depth of light extinction obviously varies with season, time of day, and cloud cover, comparison of simultaneous readings at the dump site with those at the control site will provide information on the extent to which the dumping residue reduces light transmission in the study area. Such light reduction is likely to correspondingly reduce the productivity of the area's population of phytoplankton, since phytoplankton, like all photosynthetic organisms, require light to grow and reproduce.

Recovery of the Hydrocast

"Hydrocast up!" The call from the hero platform is audible on the fantail. Now the real work begins. As the CTD is secured to the bulkhead in the "wet" laboratory adjacent to the hero platform, several scientists converge on it simultaneously. Even casual observation reveals that this is not randomly chaotic activity. Each scientist has a specific job, and all are sufficiently experienced and well briefed that the overall operation is a model of teamwork and efficiency.

One oceanographer records the temperatures of each of the reversing thermometers, using a magnifying glass as an aid. Another draws off water samples to determine dissolved oxygen content by opening a small valve on each plastic bottle and allowing the water to flow into a series of labeled glass bottles. Taking care to gently overflow the bottles (to eliminate air bubbles) and securely capping them with matching ground-glass stoppers, the scientist will then retire to the chemistry lab to carry out chemical analyses.

Two other oceanographers draw off water samples to assess concentrations of nutrients and heavy metals. The nutrients are the dissolved fertilizers of the sea. Like garden fertilizers, they provide the nitrogen, phosphorus, and other elements required for the successful growth and reproduction of marine phytoplankton and larger plants.

Nutrients most likely to be in short supply in the ocean are the nitrogen-containing compounds nitrate and ammonia and the phosphorus compound phosphate. Previous studies of this area have

FIGURE 2.10. Launching a dredge from an A-frame. (Courtesy of Woods Hole Oceanographic Institution. Photograph by Margaret Sulanowski.)

shown that, in comparison with the control site, the sewage-sludge disposal region tends to be enriched in these compounds. The biological oceanographers are particularly interested in whether phytoplankton productivity is correspondingly stimulated or whether inhibitory effects of other factors associated with dumping, such as reduced light intensity or toxic compounds, offset the benefits of additional nutrients. The nutrient concentrations will be detemined by using an *autoanalyzer,* an automated instrument which injects appropriate quantities of chemical reagents into each sample. The reagents cause a color change whose intensity can be directly related to the specific nutrient concentration.

Analysis of heavy metals yields information of a different sort. Excessive concentrations of metals such as copper, mercury, and zinc are toxic to marine life and humans and are often found in large

FIGURE 2.11. Plankton net. Mesh size is much smaller than that depicted.

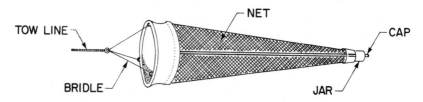

FIGURE 2.12. Zooplankton "bongo" nets being launched from R/V *Knorr*. (Courtesy of Woods Hole Oceanographic Institution. Photograph by Peter Wiebe.)

quantities in sewage sludge. When dumped into marine waters they tend to be biologically accumulated and concentrated in living organisms, so that each successive link in a food chain contains higher concentrations than the previous level. Because of the implications for fishery harvests from surrounding waters, it is important to learn about heavy metals associated with sludge dumping. The metal concentrations are determined on an instrument called an *atomic absorption spectrophotometer*.

Plankton Tows

With the CTD secured on board, the captain brings the ship to a course of 030°, magnetic compass heading, and advises the engine room that a speed of 1.5 knots (1.7 miles per hour) is required. Three biologists have gathered around a large *zooplankton* (microscopic floating animals) net on the fantail and are preparing to stream the net behind the ship. To accomplish this, they rig it to the end of a towing winch wire and reeve the wire through a block on the *A-frame*, a large pivotable boom at the stern (rear) of the vessel fashioned in the shape of a capital *A* (Figure 2.10).

The rigging secured, a crew member stands by the winch and slacks the wire as two biologists ease the net over the stern. The conical net resembles a long perforated windsock with a broad mouth, kept open by a circular steel ring, and a narrow tapering opposite end (Figures 2.11 and 2.12). The narrow end, or "cod end," of the net is also open but is securely fastened to a small plastic collecting jar. With a mesh size of 200 microns (about 0.008 inch), the net will capture all but the tiniest zooplankton (the *microzooplankton*) as it is towed slowly behind the ship. The ship will make two parallel passes, each of 5 minutes duration, in opposite directions along preselected transect lines. At the end of each pass, the net will be retrieved. Some of the contents of the collecting jar will be immediately examined; most will be preserved for later analysis.

In order to extrapolate data about population density in addition to determining zooplankton species composition, it is necessary to know the volume of water passing through the net during a tow (called the "volume filtered"). Consequently, a small flow meter has been rigged at the net entrance. As the net is pulled through the water, a propeller turns on the meter, and the number of revolutions is recorded. With this information as well as information on the duration of the tow, the distance covered, and the diameter of the net opening, the total volume filtered can be calculated. In order to know the precise depth at which the tow occurred, the scientists must make corrections for the angle which the towing wire makes with the vertical. This is determined by means of a wire-angle indicator attached to the wire.

Net sampling does not effectively recover representative samples of the smallest plankton (see Chapter 9 for a description of size classes). To adequately sample *phytoplankton* (microscopic floating plants) and microzooplankton, *Albatross IV* scientists next make a series of *Nisken bottle* casts, thereby retrieving intact water samples of known volume. They then determine species composition and abundance by later analysis of subsamples in the laboratory.

Nekton Tows

The next biota to be sampled are the *demersal* (bottom-dwelling) fish. The sampling device is an *otter trawl*, a piece of gear which is commonly used on fishing vessels and is responsible for 30 percent of the world's total fish landings. The trawl is a large net, conical in shape and tapering to a narrow end like the plankton nets, but rigged with "otter boards" on either side of the net mouth (Figure 2.13). Designed to ride in a horizontal plane when the net is towed, the otter boards serve to extend the mouth of the net and keep it open. The trawl can be weighted to ride at a desired depth. These oceanographers tow the trawl just above the bottom.

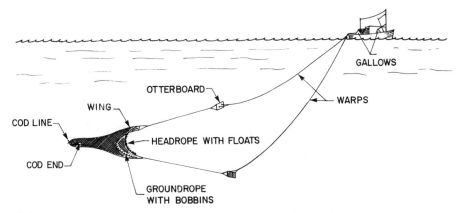

FIGURE 2.13. Otter trawl. (From Tait, R.V. 1981. *Elements of marine ecology.*
3rd ed. London: Butterworths.)

Retrieval of a trawl invariably evokes the most curiosity of any scientific activity aboard ship. Several unoccupied crew members cluster on the fantail as the net is brought aboard and emptied on the deck. The biologists work quickly to sort the species for later enumeration and identification. Unlike deep-water trawls made at depths of 500 meters or more, this catch has produced few bizarre-looking fish. Most are species commonly encountered in coastal waters of the northeastern United States in winter and primarily include hakes, cods, and flatfish.

Trawls designed for demersal fish do not retrieve nekton inhabiting the water column between the ocean surface and bottom. To obtain representative samples of these fish and invertebrates, *Albatross IV* scientists next deploy an *Isaacs-Kidd mid-water trawl.*

Sampling the Benthos

The final biological sampling is of the sea floor itself and is accomplished with two separate devices. One is an *epibenthic dredge,* constructed of a heavy steel frame to which a coarse steel net is attached. Broad, flat runners, or "skis," enable the dredge to be towed along the bottom. Thus, relatively large benthic organisms can be scooped from the mud or scraped from hard surfaces.

Epibenthic dredges tend to bounce and skip over the sea floor and are consequently rather unsatisfactory for accurate quantification of bottom organisms. A better device for this purpose is a *grab sampler,* which must be employed while the ship is stationary. Looking like a pair of hinged, open steel jaws as it plummets to the seabed on the end of a cable, the sampler can recover a known quantity of relatively undisturbed sediment and resident organisms. The "jaws" close upon contact with the bottom, thereby taking a "bite" of the seabed mud. Since the surface area and volume of sediment are known, population densities of representative organisms can be extrapolated, assuming enough replicate samples are taken.

Dredged and "grabbed" samples are carefully sieved and washed upon retrieval. As with plankton and nekton samples, captured organisms are separated and preserved for later identification and enumeration.

Respite

With the completion of the biological tows, sampling at the first station is complete. The time is 1130—in a half hour work will begin at the next station, and so the ship has begun steaming toward there now. There's time for a quick lunch, so you proceed to the mess, where about half the ship's company has already gathered. The others are on watch, working up samples, or preparing for the next round of sampling. They will eat later, after being relieved by fellow scientists or crew members. A few will skip lunch altogether, that decision dictated by their queasy stomachs.

As usual, the food is plentiful. According to custom on *Albatross IV*, the chef has prepared two separate main courses. The seas have subsided since breakfast, bringing relief to all the diners. Too frequently, meals end up on people's laps, salt shakers levitate across the room, and the simple act of drinking a cup of coffee becomes a delicate exercise.

Eating quickly, you glance around the mess. From previous cruises you know most of the scientists, who represent several different research institutions, and many of the crew members. You are inevitably struck by the camaraderie and seriousness of purpose that almost always bond the scientists aboard ship, despite extraordinarily diverse personalities, backgrounds, and interests. On short cruises personalities tend to meld; the longer the cruise, the more apparent the quirks and extremes. Usually this is beneficial, providing interest and comic relief to a routine which might otherwise become oppressive or boring.

The presence of a large proportion of graduate student scientists is a fairly recent development on oceanographic cruises. Invariably enthusiastic and hardworking, these students view the cruises

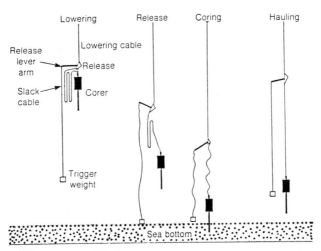

FIGURE 2.14. Principle of operation of a gravity corer. (From Duxbury, A. C., and Duxbury, A. 1984. *An introduction to the world's oceans.* Reading, Mass.: Addison-Wesley.)

as necessary apprenticeships for their chosen professions and often spend as much time at sea as they can.

Seabed Coring

After lunch you return to the main deck and confer with two geologists who are explaining the workings of a gravity-operated sediment coring device (*gravity corer*) to a student who will assist with the coring on the last day of the cruise. Basically consisting of a variable length of pipe, or "barrel," with a metal nosepiece on one end and a heavy weight and fins on the other (Figure 2.14), a gravity corer is designed to penetrate soft bottom to the length of the barrel. It is first lowered to a position just above the sea floor until a trigger weight, hanging below the corer, makes contact with the bottom. This trips a releasing mechanism on the cable, and the corer is propelled into the mud by the heavy weight of its distal end. The corer, which remains attached to the cable by a messenger wire, is then retrieved.

Once on deck, the process of removing the intact and relatively undisturbed sediment is facilitated by the presence of a plastic sleeve liner. This is readily slipped out of the barrel with the sediment inside and stored at cold temperatures, to prevent deterioration, until the sediment can be analyzed.

The afternoon goes quickly and, with the improving weather, the sampling program goes even more smoothly. Dinner comes before you know it, and then you have some free time on your hands. Free time is relatively rare on research vessels, and its disposition is a matter of personal preference. You opt to read and rest, reflecting on others you have known who have more creative interests, including one ship's master who is also a master carpenter. Since his vessel is equipped with a woodworking shop, scarcely a cruise passes which does not produce at least one beautifully wrought item of furniture or wooden toy.

Relaxing on your bunk you ponder the significance of the research conducted on this cruise. While less glamorous than the well-publicized deep-water cruises to study sea-floor spreading and continental drift or the newly discovered, teeming hot-spring communities near the midocean mountain ranges, this cruise has a purpose no less significant. By contributing even just a little more solid information on oceanic processes, whether relatively remote from human activities or deeply affected by them, we learn more about the workings of the earth we inhabit and the systems which ultimately govern our welfare and survival on this planet.

SUGGESTED READINGS

NELSON, S. B. 1971. *Oceanographic ships, fore and aft.* Washington, D.C.: U.S. Government Printing Office.

RYAN, P. R., ed. 1982. Research vessels. *Oceanus* 25(1). (A collection of articles.)

Chapter Three

The Land Beneath the Sea: An Introduction to Marine Geology

IMAGINE YOU ARE VIEWING THE EARTH FROM SPACE. IT WOULD BE clear why "Planet Ocean" might be an appropriate name for our world. Watch the globe rotate over a 24-hour period. You would see an endless expanse of water, interrupted here and there by large land masses and smaller terrestrial fragments. In fact, about two-thirds of the surface of the earth is covered by water.

The land areas which do lie on the surface of the earth are not equally distributed on the globe. A view looking down on the South Pole would reveal very little land, since more than 80 percent of the southern hemisphere is covered by water. While the northern hemisphere appears more "earthlike," land masses still occupy less than 40 percent of its surface area.

THE GLOBAL OCEAN

The dominance of water on the surface of our planet is revealed by comparisons of land elevations and oceanic depths. The average height of land is 850 meters above sea level; the average ocean depth is 3730 meters. The highest terrestrial mountain, Everest, rises over

FIGURE 3.1. Hipsographic curve indicating the proportion of the earth's surface as a function of elevation above sea level or depth of the water, Bars on left show 1000-meter frequency distributions of depths and elevations. (From Sverdrup, H.U., Johnson, M. W., and Fleming, R. H., 1942. *The Oceans.* Englewood Cliffs, N.J.: Prentice-Hall.)

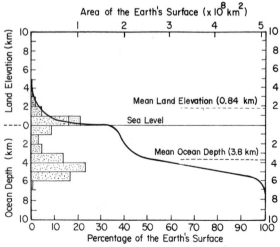

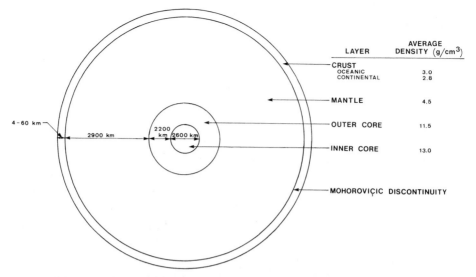

LAYER	AVERAGE DENSITY (g/cm^3)
CRUST	
OCEANIC	3.0
CONTINENTAL	2.8
MANTLE	4.5
OUTER CORE	11.5
INNER CORE	13.0
MOHOROVIÇIC DISCONTINUITY	

4 - 60 km
2900 km
2200 km
2600 km

FIGURE 3.2. Idealized crosssection of earth, showing major features. Thicknesses of various layers depicted are not proportional to actual thicknesses.

8800 meters above the sea surface, while the deepest area of the ocean floor, the Marianas Trench in the western Pacific, drops to 11,000 meters below sea level. Did you know that the highest mountain on earth lies not in the Himalayas but in the Hawaiian Island chain? Mauna Kea ascends 9763 meters from base to peak, but over half of its elevation lies underwater. And consider this. If erosion were to wear down all of the surface irregularities of the earth and uniformly redeposit the resultant sediment, a global ocean nearly two and a half kilometers deep would completely cover the planet (Figure 3.1).

THE NATURE OF THE EARTH

Despite the apparent prevalence of water on the earth, it really forms but a thin film on the surface of the earth, by one analogy being equivalent to the residual moisture left on the surface of a basketball after it is wiped with a damp cloth. Nonetheless, this liquid veneer has an importance all out of proportion to its relative mass and thickness because of its crucial role in governing atmospheric dynamics and global climate, because it comprises over 90 percent of the planet's habitable environment, and because its presence is critical to life on earth.

What materials make up the greater portion of the earth's bulk? Let us hypothetically slice the earth in half like a grapefruit. Indirect observations have revealed that the earth consists of several distinct layers (Figure 3.2). The outermost layer, bordering the atmosphere and oceans, is the *crust,* a low-density shell whose thickness, relative to the entire earth, is comparable to the relative thickness of the skin of our grapefruit. The crust of the continents consists primarily of granite, while the slightly denser crust underlying the oceans is of basaltic composition. Below the crust lies the denser, warmer, and thicker *mantle,* made primarily of magnesium and iron silicates. The crust and uppermost part of the mantle form the rigid, solid outer shell of the earth, which is commonly referred to as the *lithosphere.* Below the lithosphere lies a weak, partially molten part of the mantle known as the *asthenosphere.* The deeper, solid part of the mantle is referred to as the *mesosphere.* The innermost region of the earth is called, appropriately, the *core.* Ninety percent iron and 10 percent nickel, the core is the densest layer and the warmest and consists, we believe, of an outer liquid layer and a solid center.

Why is the earth segregated into layers? Apparently, during the planet's creation, the denser materials sank toward the center of the earth as the rotating globe cooled down, while the lighter materials migrated toward the surface of the earth.

INVESTIGATING THE INTERIOR OF THE EARTH

You might reasonably ask how we know the structure and composition of the inner earth, since despite concerted efforts we have never successfully drilled completely through the crust.

Our knowledge of the interior of the earth is indirect. While we cannot yet extract materials from deep within the earth, we can infer their composition through studies of several physical and geological phenomena. These include investigations of the behavior of sound waves within the earth, of gravity anomalies, of heat flow from the interior of the earth, of the magnetic properties of rocks, and of mantle fragments brought up in some material erupted on the surface of the earth.

The Behavior of Sound in the Earth

Scientists have known for some time that, as sound waves pass through materials, their velocity and direction are altered by the density, elasticity, and flow properties of the material.

The speed of sound waves is directly proportional to the density of the material through which they pass (the medium) and in-

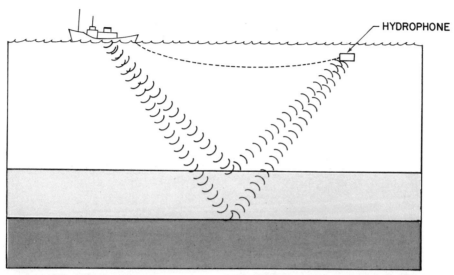

FIGURE 3.3. Seismic reflection technique. (After Gross, M. G. 1982. *Oceanography.* 3rd ed. Englewood Cliffs, N.J.: Prentice-Hall.)

versely proportional to the medium's temperature and elasticity. Thus, sound travels fastest through high-density, cold, and nondeformative materials and slowest through low-density, hot, and easily deformed materials.

The direction of sound waves is also modified by the medium through which the sound passes. A portion of the sound waves emanating from natural (e.g., earthquakes) or human (e.g., explosions) sources is *reflected* (bounced back) at the interface between materials of different densities, such as the boundaries between the different layers of the earth. But not all sound waves are reflected at boundaries; those of relatively low frequency tend to be *refracted* (bent) toward regions in which their speed is slower. The amount of refraction is proportional to the reduction of speed in the new medium.

Applying these principles, geologists use *seismic reflection* and *seismic refraction* techniques to calculate the speed and direction of travel of sound waves generated by explosions (Figures 3.3 and 3.4). These calculations then enable them to make inferences about the nature and distribution of materials within the earth. For example, seismic observations allowed scientists to recognize the boundary between the crust and mantle known as the *Mohorovičić discontinuity* (Moho for short) and to determine that its depth is 25 to 40 kilometers below the continents but only 5 to 10 kilometers below the ocean basins (Figure 3.2).

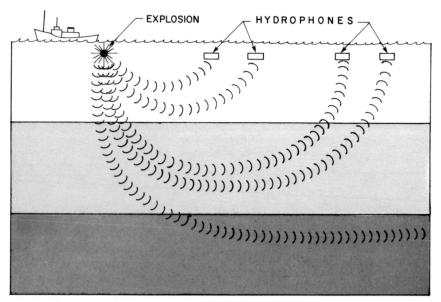

FIGURE 3.4. Seismic refraction technique. (After Gross, M. G. 1982. *Oceanography*. 3rd ed. Englewood Cliffs, N.J.: Prentice-Hall.)

Scientists have never physically penetrated the Moho, though they have tried. *Project Mohole* was an ambitious but unsuccessful attempt by oceanographers in the 1960s to drill through the oceanic crust (because it is so much thinner than the continental crust) into the underlying mantle. Although the attempt failed, the insight gained contributed significantly to the development of future ocean-floor drilling operations, including the Deep Sea Drilling Project (see Chapter 1) and offshore oil wells.

Gravity Anomalies

The composition and structure of the earth can also be inferred from variations in the gravitational field of the planet caused by unequal distribution of mass. All objects in the universe are attracted to each other by a gravitational force. The magnitude of this force is directly related to the masses of the objects and inversely related to the distance between them. The gravitational force between two bodies is consequently calculated as:

$$\text{Force} = G \frac{m_1 m_2}{r^2}$$

where G is a constant, m_1 and m_2 are the masses of the respective bodies, and r is the distance between them.

The gravitational attraction between the earth and an object on or above the surface of the earth varies because the distribution of subsurface materials varies within the earth. The denser the material directly below an attracted object, the greater the gravitational force "pulling" the object toward the earth at that location. Thus, the presence of a mass of relatively dense rock in a region of less dense material will result in a slightly higher gravity reading, or *positive gravity anomaly*, relative to the value of the surrounding region. Similarly, a *negative gravity anomaly* will be recorded where relatively light materials lie in a region of greater rock density.

Geologists routinely use extremely precise gravity-measuring instruments called *gravimeters* to detect slight deviations of gravity from the average value on earth. The extent and depth of a subsurface gravity anomaly can be calculated by knowing the density of the material causing the anomaly. Usually, a gravity anomaly is corrected for altitude (*Free-air anomaly*) or for both altitude and density differences (*Bouguer anomaly*). For example, underwater volcanoes (*seamounts*) have a positive (uncorrected) anomaly due to their topographic height. However, seamounts are composed of the same materials (basalt and sediments) as the ocean crust. Therefore, if the topographic and mass effects are subtracted, there is no anomaly.

Heat Flow from the Interior of the Earth

The interior of the earth produces a significant amount of heat, attributable in part to the residual heat from the formation of the early earth and in part to the spontaneous decay of radioactive isotopes in rocks. While the energy of internal origin received at the surface of the earth is only 0.003 of the energy received from the sun, it is still detectable. Scientists can calculate heat flow (the rate of heat loss) at different regions on the surface of the earth by inserting highly sensitive thermometers at varying depths in the crust, determining the rate of temperature increase with depth, and multiplying this value by the thermal conductivity of the rock.

In general, it has been found that heat flow from the ocean floor is roughly the same as from the continents; however, unlike the continents, heat flow from the sea floor is extremely variable. The reasons for this variability are intimately related to our evolving understanding of the interior of the earth and will be discussed in Chapter 4.

Much of what we now know about the nature, history, and origin of the geological features on earth has derived from the nascent field of *paleomagnetism*, the study of the magnetic properties of rocks. Geologists have long known that, at a high enough temperature (the *Curie point*, which is approximately 600° C), the iron particles in molten rock are free to move about in response to external magnetism. Behaving like tiny magnets themselves, the particles align themselves in the same direction as the magnetic field to which they are exposed. Once the rock cools below the Curie point, the iron particles are "locked into" position.

This behavior is immensely significant because it means that newly formed iron-containing rock, notably molten lava, will harbor, on cooling, a permanent record of the direction of the magnetic field of the earth at the time of the rock's creation. This is important because we now know that the magnetic field has not remained constant. For one thing, the apparent position of the magnetic pole has shifted in the past. This "polar wandering" has proved to constitute important evidence for continental drift (see Chapter 4). Of greater importance, however, is the astonishing recent discovery that the magnetic field of the earth, which results from the circulation of iron-rich fluid in the core, reverses polarity every half million years or so. Today, a compass needle points toward geographic north. During periods of opposite polarity, that same needle would point toward geographic south. Each reversal event takes a few thousand years to be completed. During the transition the magnetic field wanes to zero, then waxes strongly in the opposite direction where it remains until the next reversal occurs.

By determining the remnant magnetism of sea-floor rocks through towed instruments and comparing the results with the magnetic orientation of terrestrial rocks of a known age, geologists can obtain clues to help them reconstruct the history of the ocean basins. This will be discussed in greater detail in Chapter 4.

THE STRUCTURE AND COMPOSITION OF THE OCEAN FLOOR

We probably all have an innate curiosity about what lies in and on the deep-sea floor. However, our experience with the ocean bottom is usually limited to what lies below our feet at the surf zone at the

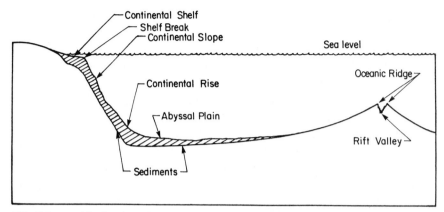

FIGURE 3.5. Idealized profile of the ocean floor. Features are not to scale and sediment thickness is exaggerated.

beach or, perhaps, the amorphous glop which clings to the weighed anchors of small boats. Some of us have donned scuba tanks and fins to extend our range of exploration. But even with diving gear, seldom do we venture deeper than 30 or 40 meters down because of high pressure, reduced visibility, and restricted communication with tenders on the surface.

But imagine for a moment that we can illuminate the abyss and that it is everywhere accessible. Further assume an inexhaustible air supply, immunity from cold and pressure effects, and a means of rapid locomotion. What would the deep-sea floor reveal to us as we skim above it on an imaginary journey across the Atlantic Ocean?

Our departure point will be a rocky outcrop on the coast of Maine, and our route will take us to the coastline of Portugal. In fact, we could have chosen any departure point, route, or ocean which struck our fancy. Though shallow-water environments vary considerably with latitude and location, the deep-sea floor, which constitutes the overwhelming portion of our journey, harbors similar features throughout the world's oceans.

The major features, more or less in order of their distance from the coastline, are the *continental margin*, consisting of the shelf, slope, and rise; the *ocean basin*, consisting of plains, volcanic mountains, hills, and valleys; and the *oceanic ridge system*. Minor but widespread features include submarine canyons, plateaus, trenches, coral reefs, and atolls. A profile of a "typical" ocean floor is shown in Figure 3.5.

THE CONTINENTAL SHELF. As we ease into the frigid waters of Casco Bay, Maine and begin our journey across the floor of the Atlantic, we may travel some 135 kilometers underwater before actually leaving the continent itself. This is because the sea floor bordering nearly all continental land masses consists of terrestrial materials or, more precisely, a currently submerged extension of the continent. These *continental shelves* as they are called, vary considerably in width. For example, the U.S. continental shelves range from 30 to 40 kilometers wide off the coast of California to more than 100 kilometers wide off the east coast. Worldwide, they average about 75 kilometers in width and occupy about 9 percent of the sea-floor area.

Continental shelves are characterized by an almost nonexistent slope (about 0.1°). Thus, their overlying waters are shallow (usually less than 135 meters deep) and, for the most part, thoroughly mixed. Thorough mixing ensures that their entire volume is exposed, at least some of the time, to the photosynthesis-stimulating rays of the sun and that nutrients are well distributed in the water column. Continental shelves are also close enough to land to receive considerable nutrient-carrying runoff from terrestrial sources. These conditions fuel a highly productive ecosystem which provides substantial fisheries resources to human populations.

There are many factors which contribute to the formation of continental shelves. One of the most important is rising sea level. Where the rising sea submerges the land, the continental shelf is correspondingly broadened. A case in point is the northeast U.S. coast. Here, after the retreat of the last extensive glaciation, sea level rose 100 to 130 meters between 18,000 and 7000 years ago (but has risen only slightly since then) as glacial meltwaters flowed into the frigid Atlantic. Today, as we glide over the teeming marine habitat in the Gulf of Maine on our imaginary journey, it is difficult to imagine that, millennia ago, pine forests once stood where stands of kelp now flourish, and herds of ungulates once roamed where schools of fish now dart in organized confusion. Yet the evidence is there for the sharp-eyed observer: here a drowned river valley once coursing with the rivulets of melting snow; there a submerged beach no longer shouldering the onslaught of storm waves. Where sea level continues to rise, still more terrestrial features will be doomed to join their predecessors below the tides, contributing their remains to the growing shelf. Conversely, should sea level drop, the shelf will shrink away, exposing once again the now-submerged relics of a former coastal landscape.

Continental shelves may form even where sea level is not rising. A storm-swept, erosional coast is a case in point. As the waves batter the shore, they eat away the rock and sand, redepositing the eroded sediment in deeper water offshore. The flat shallow terrace which results is, in fact, a continental shelf.

Shelves may also form in the absence of rising sea level if a geological disturbance causes a deformation of the coastline. Examples are the uplifting of submerged rock by earthquake activity or the eruption of underwater volcanoes. The offshore structures thus created may act as natural dams to sediments eroding from land, in turn causing the development of a continental shelf.

THE SHELF BREAK AND THE CONTINENTAL SLOPE. Gradually, our journey across the continental shelf takes us into deeper water, but the change is barely perceptible. For every kilometer traveled seaward, the water depth increases by less than 2 meters. Here, near the bottom, we are immersed in a murky void. Land-derived sediment, abundant microscopic marine life, and a host of nonliving organic materials have effectively filtered out most sunlight within a few meters of the surface. As we switch on our spotlight, we see a jumble of boulders competing for space with expanses of sediment and shapeless mounds of geological debris. The slope may be gradual, but the topography is not smooth. The mark of the ice age is evident even here. Natural hollows and crevices harbor curious eyes and waving tentacles. Less reclusive creatures, such as starfish and urchins, make pockmarks on the sandy bottom. Occasionally, the sand explodes in front of our eyes as a well-camouflaged flounder panics before our advance. The high productivity of this sea-floor environment, fed by the rain of organic debris from above, is everywhere apparent.

The bottom seems to disappear ahead. Cautiously, we inch forward. Peering into the gloom, we realize that we have reached the shelf break, where the continental shelf ends and the slope begins, some 100 kilometers from the mainland. The bottom has not really disappeared but angles more precipitously, at an average slope of 4.3°, to the deep-sea floor, 3 to 4 kilometers below. A slope of 4.3° may not seem great, but it is over 40 times as steep as the shelf and approximates the pitch of the entrance to a steep mountain road.

Carefully, we descend the slope, stabbing our light into the surrounding darkness. We have crossed an imaginary threshold. Continental slopes mark the entrance to the deep sea and, for that matter, the true ocean environment.

As rough and irregular as the shelf might have appeared, it is a monotonous plateau compared to the topography of the slope. Steep, jagged canyons cleave vertically down its incline at regular intervals, giving the slope a toothy, serrated appearance. The origin of these *submarine canyons*, whose relief and depth may rival those of the Grand Canyon, is something of a mystery. Some are obviously associated with large continental rivers, and it is not hard to imagine those canyons as former sections of riverbed inundated by rising sea level. Other canyons have no apparent connection to present-day rivers. There is indirect evidence that these canyons were formed by erosion, caused by localized downhill flows of sediment-laden water called *turbidity currents.*

While turbidity currents have never been directly observed, it is believed that they occur when a geological disturbance dislodges an accumulation of fine sediment at a shelf break. The sediment then tumbles down the slope, gathering speed as it falls and scouring the slope like a giant file. Indirect evidence for a turbidity current was obtained, quite accidentally, on November 18, 1929. On that date a major earthquake rumbled through the Grand Banks area south of Newfoundland. As the earthquake passed, an array of submarine telegraph cables, laid out along the region's continental slope, snapped in sequence from top to bottom. The last cable to break, some 13 hours after the earthquake, was located nearly 500 kilometers downslope of the earthquake's epicenter. By knowing the distance between cables and the exact time of their failures (automatically recorded by telegraph transmission machines), investigators calculated that the causative force was moving downslope at speeds of 24 to 80 kilometers per hour. Later investigation revealed a thick layer of silt lying at the base of the slope, an observation consistent with that expected from a turbidity current.

While the role of turbidity currents in creating canyons is still disputed, there is fairly good evidence that they do enlarge existing canyons. A number of observations indicate that, prior to flowing down slopes, sediment collects in the heads of canyons at the shelf break, deposited there either by transport from associated continental rivers or by longshore currents.

THE CONTINENTAL RISE. At the foot of the continental slope the steep pitch gradually begins to level off. The topography is altogether different from that of the slope, being characterized by a heavy accumulation of sediment, a very gradual pitch, and a relatively smooth surface, interrupted by shallow channels emanating from submarine canyons and small hills representing uneven sediment deposit. This transition zone between the slope and the deep-sea floor is called the *continental rise*. It is the major repository for sediments of terrestrial origin; by some estimates 40 percent of the ocean's sediments lie in this region, though it averages only 40 kilometers in width and comprises only 5 percent of the ocean floor area. While most of the sediments of the continental rise are probably deposited by turbidity currents, deep, equatorward-flowing currents along the western margins of ocean basins contribute glacially scoured sediments of polar origin.

Continental rises are conspicuously absent from a significant portion of the earth's continental margins, notably around the periphery of the Pacific basin. Since continental slopes in these regions also receive terrestrial debris, why don't sediments accumulate? The evidence is that deposited sediments are, in fact, being compacted and plastered against the continental shelf. In addition, some are actually carried into the earth's interior. The explanation for these phenomena is revealed in Chapter 4.

The Ocean Basins

Our short journey across the continental rise brings us to the true deep-sea floor. Lying at a depth averaging 4 to 5 kilometers, the deep *ocean basins* comprise 42 percent of the area of the sea floor and 30 percent of the entire earth's area.

Our first impression is of a flat, monotonous seascape. Noticeably missing are the crevices and canyons of the continental slope; even the relief of the continental shelf appears harsh in comparison. In fact, so unbroken are the features of the deep-sea floor near some continental margins that these large flat expanses are called *abyssal plains*.

Closer inspection reveals that a blanket of thick sediments is responsible for the smooth topography of the ocean basins. If we had a long enough probe, we would find that, tens of meters below the sediment surface, we would strike the hard rock of the underlying sea floor. Seismic studies have shown that this basement rock has considerable relief; however, persistent accumulations of sediments have, over the ages, filled in the crevices and valleys.

What is the source of deep-sea sediments? There is considerable evidence that most of the sediments on abyssal plains adjacent to continental margins are of terrestrial origin, and speculation centers on turbidity currents as a major agent of deposition. As we move further away from the continental rise, however, we find marked changes in the quantity and character of sediments. The depth of sediment cover becomes progressively thinner as we range further offshore, and the proportion of sediments represented by material of biological origin becomes progressively greater. Before reviewing the explanations for these observations, let's have a brief look at the nature and deposition of marine sediments.

MARINE SEDIMENTS. Sediments are the most ubiquitous and familiar of all geological features, yet they defy easy definition. Essentially consisting of unconsolidated materials, sediments may be characterized by a wide range of sizes, a variety of origins, and a host of textures and chemical characteristics.

Sediments vary in *size* from objects boulder-sized in diameter to microscopic colloidal particles. One of the most widely used sediment size scales, based on particle diameter, was devised by William Wentworth in 1922 (Figure 3.6).

Oceanic sediments originate from five major *sources,* the relative proportions of which vary with depth of water and distance from shore. *Terrigenous,* or *lithogenous,* sediments consist of eroded remains of continental rock and comprise most of the sediments of near-shore regions. A significant proportion of deep-sea sediments is of biological, or *biogenic,* origin and consists of the remains of marine organisms. Overwhelmingly, these remains are made up of the silica or calcium carbonate skeletons (*tests*) of microscopic organisms, including single-celled algae (e.g., diatoms and coccolithophorids) and amoeboid protozoans (e.g., foraminifera and radiolarians). These organisms are discussed in greater detail in Chapter 9.

The slow and steady rain of microscopic skeletons from surface waters after the death of the organisms often results in deep-sea floor sediments which are more than 30 percent biogenic. Such sediments are called *oozes.* The presence and composition of deep-sea oozes depends on the proportion of nonbiogenic sediments which

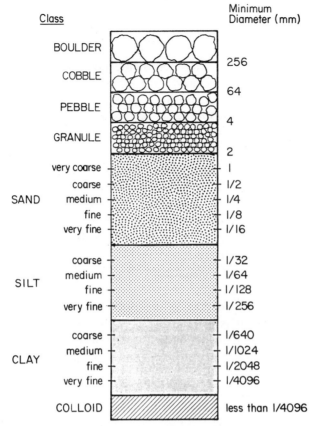

Class		Minimum Diameter (mm)
BOULDER		
		256
COBBLE		
		64
PEBBLE		
		4
GRANULE		
		2
SAND	very coarse	1
	coarse	1/2
	medium	1/4
	fine	1/8
	very fine	1/16
SILT	coarse	1/32
	medium	1/64
	fine	1/128
	very fine	1/256
CLAY	coarse	1/640
	medium	1/1024
	fine	1/2048
	very fine	1/4096
COLLOID		less than 1/4096

FIGURE 3.6. Wentworth sediment scale based on particle diameter.

dilute those of biological origin, the biological productivity of over-lying surface waters, and the chemical properties of the water, which affect the dissolution rate of the tests (see Chapter 5). The skeletons of larger marine organisms and the fecal remains of plank-tonic animals may also contribute to deep-sea sediments.

Although biogenic sediments dominate the deep-sea floor, ter-rigenous sediments are surprisingly abundant even in the abyssal depths. Fine terrestrial sediments, including products of volcanic eruptions, may be transported considerable distances by winds and currents, ultimately being deposited far from land. Coarser land-derived sediments may be rafted to the deep sea in floating ice or swept beyond the continental shelf by turbidity currents. In con-trast to the terrestrial environment, there is relatively little erosion of rock on the sea floor.

Where there are few biogenic materials on the deep-sea floor, the sediments may consist primarily of very fine inorganic particles of indeterminate origin. Such sediments are called *clays,* and their origin is thought to be related to weathering of continental rocks and/or to the alteration of marine sediments and volcanic rocks.

In isolated locales there may be a significant accumulation of mineral particles which were originally dissolved in the water. These sediments are called *hydrogenous,* or *authigenic,* because of their *in situ* origin. Examples include *evaporites* (e.g., salt deposits), which precipitate from the water column, *manganese nodules,* commercially valuable lumps of polymetallic ore which accumulate on the ocean bottom, and *metallic sulfides,* which precipitate from hydrothermal fluids emanating from the sea floor.

Finally, a surprisingly large quantity of extraterrestrial debris lands in the ocean and contributes to the seabed sediments. This *cosmogenous* material represents the remains of meteors which constantly rain down on earth. While only a few particles are large enough to be seen, it is estimated that thousands of tons of "space dust" may land on earth every day. Still, compared to other sources, cosmogenous sediments probably constitute an insignificant fraction of the oceans' total.

There are several elements of sediment *texture,* including particle diameter (discussed above.) Other elements of texture include *porosity,* a measure of the relative proportion of empty space in a volume of sediment; *permeability,* a term used to describe the relative ease with which water flows through material; and *packing,* the compactness of the grains, which in turn may affect porosity and permeability.

Knowledge of the *chemical characteristics* of sediments will reveal much about the sediments' origin, age, and suitability as a habitat for living organisms. Of particular interest to oceanographers are the relative proportions of organic and inorganic materials in marine sediments, their nutrient content, and the identity, distribution, and abundance of living marine organisms and their fossilized remains. In fact, geological and biological oceanographers routinely analyze the microscopic skeletons of marine plants and animals to help reconstruct the history of ocean basins. These techniques and their applications will be discussed in Chapter 6.

In general, with increasing distance from the continents, deep-sea sediment cover becomes thinner and increasingly of biogenic origin. The thinning of the cover is partly because, far from land, there are fewer sources of sediments, and sedimentation rates are lower. However, that is not the whole story. We have also learned that the further from land, the younger the age of the sediments, suggesting that they have not been accumulating as long as those near continental margins. The only reasonable explanation for this phenomenon is that the sea floor adjacent to continents is older than that in midocean. This surprising conclusion is discussed in detail in the next chapter.

The explanation for the increasing preponderance of biogenic sediments far from shore is more straightforward. Simply, there are fewer terrestrial sediments there. Even along continental margins, there is considerable deposition of biogenic materials. In fact, highly productive surface waters near shore produce more biological "fall-out" than offshore waters. However, the large quantities of terrigenous sediment washing into near-shore areas far exceeds, and thus masks, sediments from all other sources. In contrast, the midocean deep-sea floor is so far from continental influence that sediment sources are overwhelmingly biogenic.

RELIEF FEATURES IN OCEAN BASINS. Widespread as ocean-floor sediments may be, the deep-sea floor is not without relief. The most conspicuous of its features, which become more apparent with decreasing sediment thickness offshore, are underwater hills (*abyssal hills*), volcanic mountains (*seamounts*), *deep sea channels*, and *trenches*.

Abyssal hills are distinguished from underwater mountains by being less than about a thousand meters in height. They are ubiquitous and numerous, by most accounts being the single most abundant topographic feature on earth. Averaging about 200 meters in height, they cover approximately 80 percent of the Pacific floor and 50 percent of the Atlantic floor. Their origin is probably diverse, some apparently being the eroded remains of volcanic mountains and others representing sediment-covered elevations formed by tectonic displacement of the earth's crust.

Of greater prominence in the deep-sea floor are underwater volcanoes, or seamounts, which rise more than 1000 meters above the sea floor. While less abundant than abyssal hills, it is estimated that some 10,000 seamounts occur in the Pacific alone, with perhaps twice that many in all the world's oceans. Most are completely underwater, but some project above the sea surface, forming volcanic islands. Examples are the Hawaiian Islands, the Aleutian Islands, and Bermuda. Many seamounts whose peaks are now currently underwater were formerly volcanic islands; their great weight caused

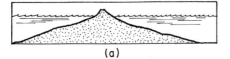

(a)

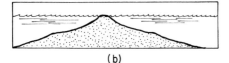

(b)

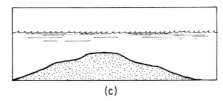

(c)

FIGURE 3.7. Subsidence of a submarine volcano: (a) Young volcanic island. (b) Subsidence has brought peak to sea level. (c) Guyot.

them and the crust beneath them to sink back into the earth through a process called *subsidence* (Figure 3.7). All volcanoes undergo some subsidence over time. Subsiding tropical volcanic islands often leave a ring of coral, called an *atoll*, surrounding their submerged peaks. The process of atoll formation will be discussed in greater detail in Chapters 11 and 12.

The peaks of seamounts which project near or above the sea surface will undergo significant erosion while exposed to waves at sea level. This will result in a flat-topped mountain, or *guyot* (sometimes called *tablemount*), after subsidence (Figure 3.7). The remains of shallow-water organisms may be found on the eroded peaks long after the volcanoes have sunk deep below the sea's surface.

As we glide across the deep-sea floor, we can see a number of shallow crevices slicing through the abyssal plains. Typically a few meters in depth, but sometimes as deep as 200 meters and up to 2 kilometers in width, these *deep-sea channels* most closely resemble river valleys. They appear to be extensions of submarine canyons and perhaps constitute a route by which terrigenous sediment is transported to the deep-sea floor, but their origin is still something of a mystery to oceanographers.

As prevalent as deep-sea channels may be, they seem decidedly inconsequential alongside the granddaddy of all marine relief features, the *ocean trench*. Plunging to depths of up to 11 kilometers, trenches represent the deepest part of the ocean environment. The *Challenger Deep*, in the Marianas Trench, is the deepest known abyss on earth, descending more than 11,000 meters below the sea surface and 4000 to 5000 meters below the surrounding sea floor.

The length of individual trenches—thousands of kilometers on average—is as impressive as their depth, but they are relatively narrow, generally less than 100 kilometers wide. Thus, their total area represents only 2 percent of the entire sea floor. Ocean trenches tend to be found in two major regions: along the border of continental land masses, and associated with chains of islands constituting what are known as *island arc systems*. Until very recently geologists had few clues as to the origin and causes of trenches. We now have considerable evidence that their formation is an important component of fundamental geological processes occurring on a global scale. A developing awareness and gradual understanding of these processes have profoundly altered our view of the earth and constitute a fascinating story of a major advancement of science over the past three decades. It is a story we will tell again here, but to tell it adequately we must first examine the last of the major submarine features we will encounter on our journey across the Atlantic.

The Oceanic Ridge System

As we continue our journey across the sea floor, a series of elevations abruptly looms ahead, almost exactly in the middle of the Atlantic. Moving closer, we are confronted by a jagged mass of peaks, ascending in steplike fashion to the murky void beyond. Scanning our underwater horizon, we realize that we have encountered not an isolated assemblage of seamounts but a true mountain range, every bit as rugged as those familiar to us on land. Were our vision unlimited, we would find that these peaks are an integral part of an enormous, globe-encircling, underwater mountain range—the *oceanic ridge system* (Figure 3.8). The total length of the system approximates 65,000 kilometers, more than one and a half times the world's circumference, and its area is nearly a third of the entire sea floor and over 23 percent of the earth's area. Altogether, the oceanic ridge system covers nearly as much of the globe's surface as all of the earth's land masses.

The mountain range which we have happened upon bisects the Atlantic Ocean from north to south and is called the *Mid-Atlantic Ridge* (Figure 3.8). It averages 1500 to 2000 kilometers in width,

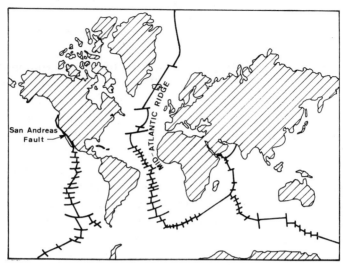

FIGURE 3.8. Locations of major features of the oceanic ridge system. Lines perpendicular to main axis indicate fracture zones. (After Gross, M. G. 1982. *Oceanography.* 3rd ed. Englewood Cliffs, N.J.: Prentice-Hall.)

and its highest elevations project nearly 3 kilometers above the surrounding sea floor, often to within 2 kilometers of the sea surface above. (In fact, in some areas of the North Atlantic, notably the Azores and Iceland, the Mid-Atlantic Ridge actually rises above the sea surface). Slicing across the ridge, deep perpendicular gashes, called *transform faults,* divide the ridge into a series of offset segments (Figure 3.9).

Despite being the single largest continuous geological feature on earth, the oceanic ridge system was completely unknown to scientists until the 1950s, when it was revealed through seismic profiling of the ocean floor. Its discovery raised some perplexing questions. What is its origin? Why is it a continuous underwater feature? Why does the Mid-Atlantic Ridge exactly bisect the Atlantic Ocean, and as was subsequently discovered, why do the features of the sea floor appear so symmetrical on either side of the ridge? As scientists made a concerted and cooperative effort to study the oceanic ridge system, the answers to these questions were not long in coming. But before revealing what we now know about the oceanic ridge system and its critical role in global geology, let us continue our journey across the Mid-Atlantic Ridge, where we will encounter one of the most important and spectacular components—the *rift valley.*

RIFT VALLEYS. Carefully, we work our way up the rugged topography of the Mid-Atlantic Ridge until we have ascended its highest peaks, located on the ridge axis. Pausing for breath, we survey the scene around us. Behind us, toward the North American coast, waves of jagged pinnacles cascade down to the flat abyssal plains faintly visible in the distance. Before us, toward Europe, an even more remarkable view greets us. Our peak descends steeply into a deep chasm in whose depths, 1 or 2 kilometers below, we might see (if our segment of ridge were currently active) the molten glow of lava and scattered geyserlike jets of turbid water.

The chasm before us is a rift valley, a common (but not universal) feature of the oceanic ridge. Cleaving the ridge axis, the rift valley averages only 25 to 50 kilometers in width and is bordered, on either side, by the highest peaks of the ridge. On the ridge crest, as well as in the valley, there is evidence of strong tectonic activity. The ridge peaks are clearly of volcanic origin, and the fissures of molten lava deep in the rift valley also indicate persistent plutonic disturbance. Volcanism is not the only manifestation of a deeply disturbed earth here. Immense fracture zones (*transform faults*) perpendicular to the ridge axis indicate pronounced earthquake activity. Even now, as we perch warily on a peak, the ridge rumbles ominously below us.

Controlling our impulse to flee, we carefully observe more details of our surroundings. One immediate discovery is the marked

FIGURE 3.9. Transform faults between displaced sections of oceanic ridge. (a) 1 = segments of Plate 1; 2 = segments of Plate 2. Rift at ridge peak separates segments of different plates. (a) and (b) Arrows indicate direction of plate movement. (b) Earthquake activity is high where segments of adjacent plates move in opposite directions.

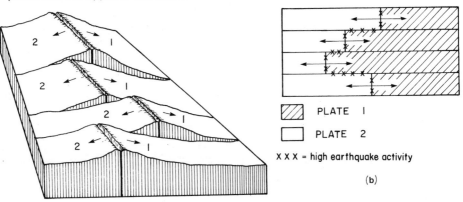

PLATE 1

PLATE 2

X X X = high earthquake activity

(b)

(a)

paucity of sediments on the oceanic ridge. On our journey across the sea floor we have been aware that the thickness of the sediments has been steadily decreasing with distance from shore. Here on the ridge only a thin covering of sediments dusts the crags and crevices. Now, as we move off our pinnacle vantage point and swim across the gulf separating us from the line of peaks on the other side of the rift valley, we make another important observation: The topography on opposite sides of the rift valley is strikingly symmetrical. Not only are the ridge inclines and gradually decreasing elevations remarkably similar, but the patterns of sediment thicknesses are also mirror images.

Unknown to us, without sophisticated instrumentation, is the fact that this symmetry is even more subtle and profound than is apparent here and extends to the shores of the continents themselves, on opposite sides of the ocean. Evolving awareness of this symmetry and its underlying causes have, more than any other single factor, enabled scientists to confirm the once implausible notion of continental drift and led to the development of a globally unifying concept in geology: *plate tectonics.*

SUGGESTED READINGS

Cox, A., Doell, R. R., and Dalrymple, G. B. 1964. Reversals of the earth's magnetic field. *Science* 144:1517-43.

Emery, K. O. 1969. The continental shelves. *Scientific American* 221:106-22.

Heezen, B. C. 1956. The origin of submarine canyons. *Scientific American* 195:36-41.

Heezen, B. C., and Hollister, C. D. 1971. *The face of the deep.* New York and London: Oxford University Press.

Mark, K. 1976. Coral reefs, seamounts, and guyots. *Sea Frontiers* 22:143-49.

Shepard, F. P. 1973. *Submarine geology.* 3rd ed. New York: Harper and Row.

Spencer, D. W., Honjo, S. and Brewer, P. G. 1978. Particles and particle fluxes in the ocean. *Oceanus* 21:20-26.

Chapter Four

Plate Tectonics and Sea-Floor Spreading

IN THE HARSH AND FRIGID LANDSCAPE OF WESTERN GREENLAND lie the oldest rocks known on earth. Sculpted and smoothed by the blasts of a million storms, they perch in mute wisdom and infinite patience, like ancient sentinels in a forbidden land.

There is something comforting in those rocks. Nearly 4 billion years old, they have seen over four-fifths of the history of the earth. In a world of human chaos, with its uncertain future and frequently indecipherable past, they represent stability, continuity, and survival. We appreciate their persistence.

Still, these rocks are geological oddities, anachronisms on an earth where the norm is a quick passage to obscurity. Though we may perceive an essentially immutable landscape, in reality the surface features of the earth are ephemeral and, for the most part, surprisingly young.

Sometimes we have a sense of this instability. We acknowledge the gradually destructive effects of erosion. More disquietingly, the earth periodically belches and rumbles, signifying a sort of global gastric distress. Occasionally, this geo-indigestion produces catastrophes in the form of violent earthquakes and volcanic eruptions. The earth is an active place.

How widespread are these disturbances and of what global significance? Are they chance unrelated occurrences or part of an intricate global pattern? Several startling discoveries over the past three decades have begun to provide answers to these questions. The discoveries have largely derived from oceanographic investigations and have radically changed our understanding of the fundamental workings of the earth. The description of these discoveries and their global significance constitutes the essence of this chapter. Let us first, however, make a slight historical digression.

ALFRED WEGENER AND THE THEORY OF CONTINENTAL DRIFT

In 1915, a brilliant German meteorologist, astronomer, and explorer named Alfred Wegener published a remarkable book. *The Origin of Continents and Oceans* brought together a wealth of fossil, geological, and climatological data to support Wegener's hypothesis that the separate continents on earth were once united. Wegener called this proposed supercontinent *Pangaea* and theorized that about 300 million years ago it split into fragments that subsequently drifted

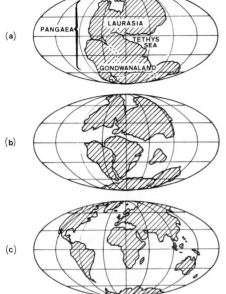

(a)

(b)

(c)

FIGURE 4.1. The movement of the earth's continents since the breakup of Pangaea. (a) 225 million years ago. (b) 135 million years ago. (c) Today. (After Duxbury, A. C. and Duxbury, A. 1984. *An introduction to the world's oceans.* Reading, Mass.: Addison-Wesley.)

apart (Figure 4.1). These fragments, argued Wegener, became today's continents, even now drifting on the earth's surface. The concept of major land masses sliding around the earth like checkers on a board was preposterous to most early twentieth-century readers, but Wegener presented some impressive supporting evidence.

He first reaffirmed the remarkable geometrical compatibility of the terrestrial borders of the continents surrounding the Atlantic Basin. This was not a new observation, having first been formally noted by Francis Bacon in his *Novum Organum*, published in 1620. Wegener, however, presented much more evidence, which was largely codified from findings of several earlier investigators. This included the surprising similarity of ancient geological and paleontological features on opposite sides of the Atlantic, including the regions' mineral compositions, the structure and ages of mountain ranges, fossil remains, and ancient climates. It is highly unlikely that such similar features evolved simultaneously and independently in geographically distinct regions, separated by expanses of ocean. More likely, in Wegener's opinion, the land masses were once one.

Wegener's *continental drift hypothesis* was plausible and interesting, but it also was revolutionary, long on speculation and short on hard data, and failed to propose a feasible mechanism for the movement of continents. Not surprisingly, the hypothesis failed to garner lasting support, despite considerable interest at the time. By the 1930s Wegener's ideas were all but forgotten.

CONTINENTAL DRIFT RECONSIDERED: THE MODERN GLOBAL GEOLOGY

The 1950s and 1960s saw the rebirth and expansion of Wegener's continental drift hypothesis. The renewed interest stemmed from significant new evidence derived from studies of the topography, heat flow, age, and magnetic properties of the ocean floor.

Ocean Floor Topography

We have already talked about the enormously important discovery of the oceanic ridge system and associated rift valleys. We noted the striking similarity of geological features on opposite sides of the ridge axis. This discovery, coupled with the long-recognized fit of continents surrounding the Atlantic Basin, lent further credence to the growing feeling that this symmetry might not be coincidental. Furthermore, evidence of strong volcanic and earthquake activity associated with the ridge system suggested something was happening there that demanded explanation.

Heat Flow

We have also referred to the discovery, again in the 1950s, that the heat flow emanating from the ocean floor is neither uniform nor of random variability. In fact, sensitive thermometers showed that the highest flow occurs at the oceanic ridges, that heat flux declines symmetrically toward the continents on opposite sides of the ridges, and that it is lowest adjacent to continental margins. Whether this pattern was attributable to geographical differences in the magnitude of the heat source, to relative differences in the distance of the heat source from the sea floor, or to other unspecified factors was an intriguing question which taxed the imaginations of geologists in the 1950s.

Age of Crustal Rocks

The 1950s also saw the first accurate dating of crustal rocks on the sea floor. We may recall that the oldest rocks found on land were formed nearly 4 billion years ago. Astonishingly, it was found that most ocean-floor rock was less than 100 million years old and that no seabed crust had existed for more than 200 million years. This implied the seabed was less than one-twentieth the age of the continents. Most scientists agree that the oceans' waters have been around a long time—since shortly after the origin of the earth itself. Why, then, is the sea floor so young?

The Hypothesis of H.H. Hess

In 1960 Harry H. Hess, a geologist from Princeton University, presented an intriguing paper at a scientific meeting. Entitled "History of the Ocean Basins," the paper drew largely on the discoveries of the 1950s. Hess hypothesized that the solid lithosphere of the earth is broken into large plates which move around on the semifluid asthenosphere in response to convection cells created by heat emanating from the interior of the planet. How do convection cells work? Consider the behavior of blocks of styrofoam floating in a beaker of water heated from below (Figure 4.2). As the water closest to the heat source warms and expands, its reduced density causes it to rise. On reaching the surface it spreads out, carrying the styrofoam with it, then cools, becomes denser, and sinks again. The cycle is repeated when the sinking water approaches the heat source again. The re-

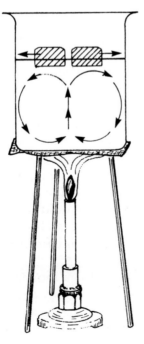

FIGURE 4.2. Convection cells in a heated beaker of water.

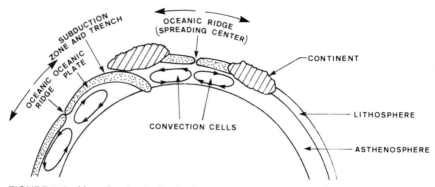

FIGURE 4.3. Hypothesized effect of convection cells on movement of segments of the earth's lithosphere. Arrows indicate directions of segments' movement.

sultant circular vertical current is called a *convection cell*. In a similar fashion, speculated Hess, segments of lithosphere, with their embedded continents, would be carried across the surface of the earth in response to thermal convection cells in the asthenosphere.

Hess proposed that heated material from the asthenosphere ascended to the surface of the earth at oceanic ridges. He based this conclusion on evidence of abnormally high heat flow and volcanic activity in these regions. Hess theorized that molten material oozing out of rift valleys would provide a constant source of new ocean floor upon cooling and solidifying.

An unceasing supply of new ocean-floor material might suggest a constantly expanding earth. However, there is no evidence that the earth is getting larger. Alternatively, Hess proposed that descending portions of the asthenosphere convection cell would drag portions of the sea floor down with it (Figure 4.3). Hess called this region of sinking a *subduction zone* and formulated the term *plate* to describe a segment of lithosphere bounded by an oceanic ridge on one side and a subduction zone on the other. Hess proposed that the deep ocean trenches are locations of subduction zones and that the trenches are created when the leading edge of one moving plate descends below another converging plate (Figures 4.3 and 4.4). The constant creation and destruction of ocean floor would consequently mean that existing seabed is everywhere youthful, compared to the continents which largely remain on the surface of the earth.

Hess's hypothesis provided a rational and verifiable explanation for a number of recent, previously irreconcilable geological discoveries. It was certainly speculative and deficient in supporting data, leading Hess himself to refer to it as "geopoetry." However, unlike Wegener's hypothesis five decades earlier, Hess's ideas caught hold quickly and prompted a flurry of scientific activity designed to test their validity. The supporting evidence was not long in coming.

Supporting Paleomagnetic Evidence

Supporting evidence for Hess's sensational hypothesis came from paleomagnetic observations of the seabed. Recall from Chapter 3 the finding that the magnetic field on earth has periodically reversed polarity. This conclusion derived from a curious oceanographic discovery in the early 1960s. When towing highly sensitive magnetometers above the seabed in an attempt to chart *magnetic anomalies* (slight deviations in the magnetic field), geologists found that seabed rocks in the northeast Pacific Ocean displayed parallel bands of alternating magnetic polarity. Long, narrow (about 30 kilometers wide) strips of rock with northward-pointing polarity were bordered, on either side, by strips of opposite polarity. In 1963 two British geophysicists, F.J. Vine and D.H. Matthews, concluded that this showed that the entire earth has undergone periodic magnetic reversals throughout its history and that sea-floor rock simply reflects the prevailing polarity at the time of the rock's formation from molten materials.

FIGURE 4.4. Subduction zone at the leading edges of converging oceanic and continental plates.

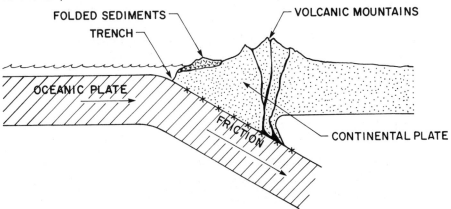

FIGURE 4.5. Sediment cores recovered from a depth of over 4500 meters. Each core has been divided lengthwise. One half is used for geological, chemical, and biological analyses. The other half is stored at low temperature to provide a permanent record. (Courtesy of Woods Hole Oceanographic Institution. Photograph by Al Driscoll.)

The application of paleomagnetic techniques to the study of ocean-floor history was extended to investigations of sediments. Magnetic particles in sediments also align themselves in the direction of the magnetic field of the earth as they settle to the sea floor. The net magnetism of these particles is much weaker than that of solid rock but is still detectable. Studying sediment magnetism is in some ways preferable to studies of rock, since sediments provide a longer and more continuous record of polarity than rock, which is of intermittent volcanic origin. Thus a long, narrow tube of intact sediment recovered by a coring device (Figure 4.5) can provide a record of the earth's magnetism over 100 million years or more, with the most recent polarity recorded at the sediment surface.

Accurate dating of magnetic events recorded in the sediments is accomplished in several ways. Where sediment settling rate can be calculated for long periods, the age of the sediments can be correlated with its depth. Also, to some extent, the mineral components of the sediments can be dated radioactively. Finally, fossilized marine skeletons in sediments can be accurately dated, using techniques described in Chapter 6. Their age can then be correlated with sediment depth.

In contrast, nonsedimentary ocean-floor rock can only be accurately dated to about 5 million years. This is accomplished primarily by comparing the pattern of magnetic strips on the sea floor

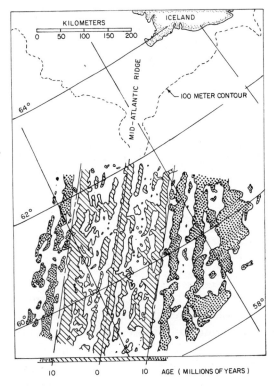

FIGURE 4.6. Depiction of orientation of
magnetic field displayed by sea floor rocks
adjacent to Mid-Atlantic Ridge south of
Iceland. Ridge axis is at "Age O." Youngest
rocks (0 to 10 million years old) display
"normal" polarity. Older rocks (greater than
10 million years old) display "reversed"
polarity. Polarity is "locked in" on cooling of
molten rock when sea-floor crust was
formed. A constant sea-floor spreading rate
is assumed. Note symmetry of magnetic
patterns on opposite sides of ridge axis.
(From Gross, M. G. 1982. *Oceanography.*
3rd ed. Englewood Cliffs, N.J.: Prentice-Hall.)

with the more easily assessed patterns of geologically recent terrestrial lava flows.

The important supporting evidence for Hess's hypothesis came from studies of magnetic polarity recorded in seabed rocks and sediments on opposite sides of oceanic ridges. Not only were alternating bands of polarity found, but it was also discovered that the pattern of polarity reversals was strikingly symmetrical about the ridge axis. The same band sequences and widths were evident on both sides. Such a pattern is exemplified by ocean floor bordering the Mid-Atlantic Ridge (Figure 4.6).

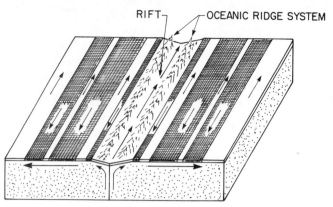

FIGURE 4.7. Model of sea-floor formation and spreading. New floor is created at rift valley. Oldest floor is furthest from ridge axis. Arrows on bands indicate direction of magnetic polarity at time of sea-floor formation.

Suddenly, pieces of the puzzle began to fall into place, and Hess's proposed model of sea-floor spreading began to look accurate. If new sea floor is created at midocean ridges from molten magma of asthenosphere origin, and if it subsequently moves away from both sides of the ridge axis at the same rate, as if carried by a dual conveyor belt, then a symmetry of magnetic polarity bands would indeed be displayed. Such a model is depicted in Figure 4.7. According to this model, the oldest portions of the sea floor would eventually be carried back down into the earth as the cooling asthenosphere again descends, accounting for the fact that the sea floor is nowhere very ancient.

To provide further verification of this model of *sea-floor spreading* and to couple it with the concept of plate tectonics, it was next necessary for oceanographers and geologists to focus their observations on the plate boundaries: those regions where ocean crust originates and where it is subsequently destroyed.

Plates, Plate Movement, and Plate Boundaries

Scientists now believe that the lithosphere is broken up into seven major plates and a number of minor plates, each moving independently of each other on the lubricating asthenosphere layer (Figures 4.8 and 4.9). Three kinds of relative plate movement are evident at plate boundaries: *divergent* movement, where plates are spreading apart; *convergent* movement, where plates are subducting; and *lateral* movement, where plates slide past each other, as at transform-fault boundaries (Figure 4.9). Each boundary type is characterized

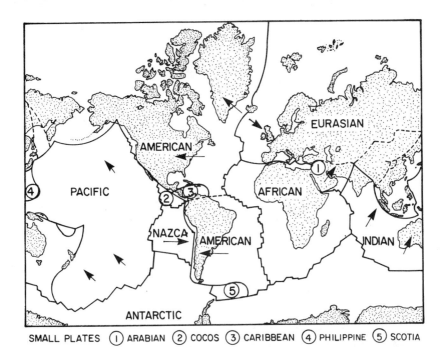

SMALL PLATES ① ARABIAN ② COCOS ③ CARIBBEAN ④ PHILIPPINE ⑤ SCOTIA

FIGURE 4.8. The earth's major lithospheric plates. Arrows indicate direction of plate movement. (From Duxbury, A. C. and Duxbury, A. 1984. *An introduction to the world's oceans.* Reading, Mass.: Addison-Wesley.)

FIGURE 4.9. Major lithospheric plates of the world showing types of plate boundaries. (Courtesy of Woods Hole Oceanographic Institution.)

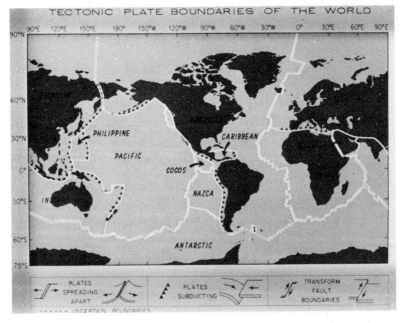

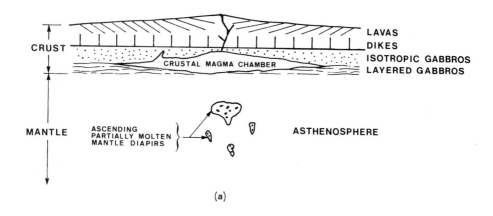

FAST SPREADING RIDGE
(e.g. EAST PACIFIC RISE)

CRUST

LAVAS
DIKES
ISOTROPIC GABBROS
LAYERED GABBROS

CRUSTAL MAGMA CHAMBER

MANTLE

ASCENDING
PARTIALLY MOLTEN
MANTLE DIAPIRS

ASTHENOSPHERE

(a)

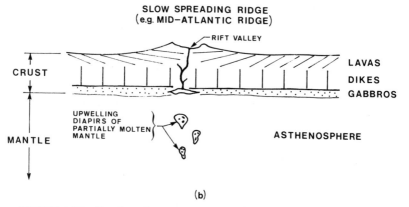

SLOW SPREADING RIDGE
(e.g. MID–ATLANTIC RIDGE)

RIFT VALLEY

CRUST

LAVAS
DIKES
GABBROS

MANTLE

UPWELLING
DIAPIRS OF
PARTIALLY MOLTEN
MANTLE

ASTHENOSPHERE

(b)

FIGURE 4.10. Sea-floor formation at oceanic ridges. (Interpretation by P. Meyer, Woods Hole Oceanographic Institution.)

by considerable earthquake and volcanic activity. Let's summarize what we now know about these geologically active regions of the earth.

DIVERGENT PLATE BOUNDARIES. Divergent plate boundaries are located at oceanic ridges. Recall the environment of an oceanic ridge system. It is a violent place marked by jagged peaks and chasms, the occasional rumbling of earthquakes, and periodic eruptions of molten rock. The rift axis is the source of new sea floor, born of hot, low-density lava rising from chambers of magma located within and at the base of the crust. These magma chambers are fed from upwelling *diapirs* (plumes) of partially molten mantle (Figure 4.10).

The rising magma creates the elevated rock formations which characterize the oceanic ridges. Later, the completely cooled lava spreads away from the ridge axis in the form of new oceanic crust, and the jagged topography is eventually smoothed by erosion and sedimentation. At the rift valley, additional magma intrusions fill the void left behind. Thus, the ocean floor is born and renewed.

Fissures may form in the cooling lava above the magma chambers. The cold ocean water which flows into these cracks heats up to temperatures as high as 350° C, subsequently exploding back out in a geyserlike eruption. These *hydrothermal vents* have only recently been discovered but have generated intense interest. They are sometimes the sites of enormously productive animal communities whose existence, far below the sun's life-giving rays, initially baffled oceanographers (see Chapter 10). They also appear to have a significant effect on sea-water chemistry by contributing large quantities of new minerals, created by the chemical and thermal interactions of cold ocean water with hot magma, to the oceans' waters. The number and total area of thermal vents is apparently not large, but because of high mineral concentrations their contribution to ocean chemistry may be equal to that of all continental rivers.

The crest of oceanic ridges is not a continuous, unbroken feature but is frequently offset along its axis by transform faults (Figure 3.9). Earthquakes often occur along fault lines because sections of adjacent sea floor are moving in opposite directions with considerable frictional resistance.

CONVERGENT PLATE BOUNDARIES. If segments of the earth's lithosphere are separating as distinct plates at oceanic ridges, then elsewhere plates must be colliding together on our spherical globe. Convergent plate boundaries occur where the leading edges of two different plates meet (Figures 4.3, 4.4, and 4.9). Despite the relatively slow rate of convergence (characteristically a few centimeters a year), resultant forces are tremendous. The earth groans and shudders at convergent plate boundaries, and the face of the planet is transformed in the process. Here we find the earth's highest mountains and its deepest trenches. Wrenching earthquakes and explosive volcanoes vie in their assault on the landscape. It is a place of violence and extremes.

What happens when plates converge? If a plate consisting exclusively of oceanic crust, exemplified by the massive Pacific Plate, drives into a plate with a continent at its leading edge, the oceanic

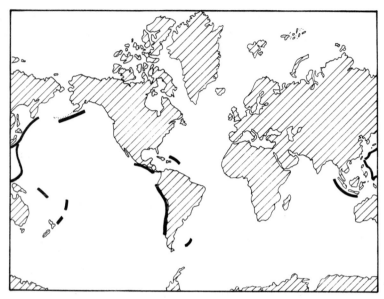

FIGURE 4.11. Location of oceanic trenches (indicated by dark bands). (From Duxbury, A. C. and Duxbury, A. 1984. *An introduction to the world's oceans*. Reading, Mass.: Addison-Wesley.)

plate is subducted below the continental plate because the basaltic oceanic crust is slightly denser than the granitic continental crust (Figure 4.4). A deep trench, formed when the leading edge of the oceanic plate plunges into the asthenosphere, characteristically occurs on the seaward side of the boundary. On the landward side a mountain range is created by the folding and uplifting of continental shelf sediments when the plates grind together (Figure 4.4). An example is the convergence of the eastern margin of the Nazca Plate with the western boundary of the South American Plate, with the continent of South America embedded at its leading edge (Figure 4.8 and 4.9). The deep Peru-Chile Trench lies immediately offshore and is paralleled by the Andes mountain range, about 300 kilometers inland.

Ocean trenches are not confined to continental margins (Figure 4.11). Throughout the western Pacific, the Pacific Plate converges with several eastward- or southward-moving plates whose leading edges are several hundred kilometers distant from continental land masses. On the plates closest to the Asian mainland lines of volcanic islands, called *island arc systems*, erupt at the plate boundary while just seaward some of the oceans' deepest trenches occur.

Shallow seas lie between the Asian mainland and the island arc system, and land-derived sediments overlie the ocean floor. The Aleutian Islands and the islands of Japan and Indonesia are examples of island arc systems.

Frictional contact at the interface of converging plates results in strong earthquake activity, which can be traced along the plate margins almost to the depth of the asthenosphere. These earthquakes typically have a deeper average center of activity or "focus" (100 to 200 kilometers) than those at oceanic ridges and show a pattern of increasing depth toward the land. Resultant friction and inner-earth heat cause partial melting of descending ocean crust and/or the overlying mantle wedge. Melting in these regions is greatly enhanced by water that is released from the subducted, hydrated ocean crust. The high concentration of water and other volatiles in magmas generated in this environment results in violently explosive eruptions, recently exemplified by the catastrophic 1980 eruption of Mount St. Helens in the Cascade Range of the state of Washington. Explosive eruptions such as this occur all along convergent plate boundaries and are characteristic of the perimeter of the Pacific Ocean (Figure 4.9). Because of intense volcanic activity, the perimeter of the Pacific Basin is often referred to as the "Ring of Fire."

Volcanic eruptions along convergent plate boundaries generate large volumes of *tephra* (airborne fragments of volcanic rock), which form important stratigraphic markers in marine sediments. The volcanoes themselves are often composed of both tephra layers and lava flows and are called either *stratovolcanoes* or *composite volcanoes*. Tephra and lava produced in these volcanoes are generally far more siliceous than basalt, ranging in composition from andesite to rhyolite, the fine-grained equivalent of granite. Thus, volcanism along convergent plate boundaries is markedly different from the comparatively passive eruption of basaltic lavas at oceanic ridge systems.

The mountainous Hawaiian Islands are also of volcanic origin, but they are exceedingly distant from plate boundaries and not associated with island arc systems. What, then, is their origin?

If we examine a map, we clearly see that the islands form a long chain stretching from southeast to northwest. If we then date the rocks on the various islands, we find a curious result: The islands are progressively older toward the northwest, and the most recently formed islands lie farthest to the southeast. Even more surprising, the increase of age with distance toward the northwest is closely correlated with the calculated rate of sea-floor spreading at the East Pacific Rise (the spreading center which forms the boundary between the Pacific and Nazca Plates; Figures 4.8 and 4.9). Consequently, geologists have tentatively concluded that, as the Pacific Plate moves toward the northwest, it slides over a stationary area in the mantle called a *hot spot*, where thermally buoyant plumes or diapirs are generated.

As these plumes rise, the entrained (incorporated) mantle material begins to melt. Periodically, this activity produces enough magma to fuel volcanic eruptions on the sea floor. The result is an archipelago of volcanic islands whose orientation reflects the direction of plate movement (Figure 4.12). Other Pacific island chains of apparently similar origin include the Marshall Islands, the Line Islands, and French Polynesia.

Occasionally, two separate continents converge at plate boundaries. Since the crustal materials are of similar density, the continental crust is apparently not subducted. Instead, the converging plate boundaries collide, pushing up shelf rock and sediments into continental mountain ranges. Because the materials of these mountains originate largely from the former beds of shallow coastal seas, fossil

FIGURE 4.12. Hypothesized formation of volcanic island chains by "hot spots." (Adapted from Stowe, K. 1983. *Ocean science.* 2nd ed. N.Y.: Wiley.)

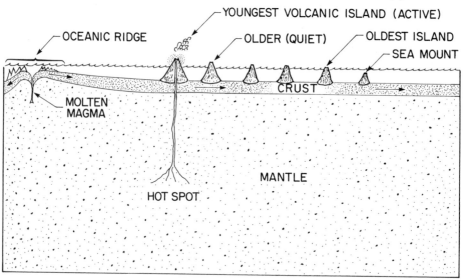

remains of ancient marine organisms are frequently found in the mountains' sedimentary strata.

The Himalayan mountain range is the most spectacular example of the uplifting of continental margin materials into a major mountain range. Nearly 200 million years ago the present subcontinent of India was separated from modern-day Asia by an expanse of ocean (Figure 4.1). The geological evidence indicates that by 75 million years ago the Indian subcontinent, embedded within the Indian Plate, was moving northeasterly toward the plate embodying the Asian mainland. Perhaps 35 million years ago the two plates collided. Initially, the oceanic crust on the leading edge of the Indian Plate descended into the asthenosphere. As the land masses continued to move together, the merging continental margins were uplifted, becoming the Himalayan mountain range. These land masses press against each other even today. As they do so, the Himalayas continue to grow.

Most of the major mountain ranges on earth resulted from such collisions of continents. In some cases, where the land masses subsequently separated again, the mountains are gradually being whittled down through erosion. For example, the Appalachian Range in the eastern United States and the mountains of western Europe were apparently formed simultaneously when, several hundred million years ago, the continents of Europe and North America converged. After the continents subsequently separated at the Mid-Atlantic Ridge, the mountain range was cleaved and both halves have been eroding ever since.

TRANSFORM FAULTS. There are several areas on earth where plates are neither converging nor separating but are sliding past each other. Because of their rough, irregular boundaries, the adjacent plates may temporarily hang up on each other. Pressure builds up. Eventually, the plates break free; with the sudden release of pressure, earthquakes rend the earth. Their magnitude depends directly on the amount of prior pressure buildup.

The most widely cited example of such a plate boundary is the San Andreas Fault in California. On the west side of the fault a large segment of California, which includes Los Angeles County, is slowly sliding toward the northwest on the Pacific Plate (Figure 4.8). On the east side of the fault the remainder of California, including San Francisco, moves westward with the rest of the North American Plate. Thus, the net motion of the adjacent plates is lat-

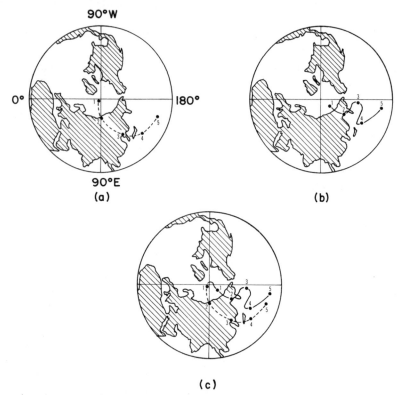

FIGURE 4.13. Positions of magnetic North Pole ("polar wandering curves") as determined from: (a) European rocks; (b) North American rocks. (c) Curves from each continent are superimposed. Numbers are hundreds of millions of years ago. (After Weyl, P. 1970. *Oceanography: an introduction to the marine environment.* N.Y.: Wiley.)

eral, and Los Angeles slips slowly but inexorably toward San Francisco. The earthquakes which occur when the plates hang up and subsequently break apart are well documented, and there is a consensus among geologists that future catastrophes are inevitable.

Supporting Evidence from the Continents

The concepts of plate tectonics and sea-floor spreading are further supported by paleomagnetic and structural studies of continental land masses conducted during the 1950s and 1960s. These included investigations of apparent deviations in the location of the earth's magnetic pole (termed "polar wandering") and precise determinations of the boundaries of continental margins.

POLAR WANDERING. Recall from Chapter 3 that the earth's magnetic pole apparently "wanders" in addition to its periodic reversal of orientation. In order to determine whether, over time, the position of the magnetic pole actually changes or whether the continents have moved relative to the earth's magnetic field, investigators deliberately made the false assumption that the continents have remained stationary. They then plotted apparent past positions of the magnetic North Pole, based on the calculated magnetic orientation of rocks in Europe and North America over the past 500 million years (Figure 4.13). This allowed them to draw two "polar wandering" curves from the separate perspectives of the two continents. If neither continent had moved, the independently derived curves should have been precisely superimposable. In fact, they were not (Figure 4.13). The only credible explanation is that one or both continents have indeed moved, substantiating the hypothesis of continental drift.

CONTINENTAL MARGINS. The other evidence from the continents came from advances in sea-floor mapping which were made possible by improved seismic survey techniques in the midtwentieth century. Once the outlines of the continental shelves surrounding the Atlantic Basin were precisely determined, it was evident that the "fit" of continents on opposite sides of the ocean was even more compatible along the continental shelf edges than along the present terrestrial borders. The English geophysicist, Sir Edward Bullard, constructed a computer fit of the continents in 1965. He found an astonishing match of continental borders along the 2000-meter-depth contour, about halfway down the continental slope, rendering even less likely the possibility that the geometrical similarity of the continents was strictly coincidental.

Causes of Sea-Floor Spreading

While there is little doubt today of the existence of continental drift and sea-floor spreading, there is still a great deal of uncertainty as to their causes. Hess's hypothesis of convection cells within the mantle continues to be the most popular, but many puzzling questions remain unanswered. For example, what is the scale of the cells, and is the circular motion of the cells capable of driving plate tectonics? Some scientists believe that the plates are completely decoupled from the mantle due to the partially molten nature of the asthenosphere and advocate that the driving force for plate motion lies within the plates themselves. Other forces that may be driving plate tectonics include the pulling power of subducting lithosphere at trenches and the pushing force of rising asthenosphere at oceanic ridges.

There are other intriguing questions. What is the origin of oceanic ridge systems, and how long-lived are they? What phenomena are responsible for the apparent alternating expansion and contraction of some ocean basins, exemplified by the Atlantic? Why is the rate of sea-floor spreading some five times as great in some parts of the world's oceans as in others? In addition to uncertainty regarding the mechanisms of sea-floor spreading, the history of the ocean basins is also largely a mystery—but one which a host of oceanographers is currently attempting to resolve. The following section summarizes what we know to date.

A BRIEF HISTORY OF THE OCEAN BASINS

We stated earlier that no part of the sea floor is more than about 200 million years old. The ocean bottom is everywhere youthful, compared to the continents, because of the continuously creative and destructive processes of sea-floor spreading and plate tectonics. Consequently, we have only been able to reconstruct the past 220 million years of ocean-basin history—only about 5 percent of the earth's entire lifetime. However, even within that relatively short time span, the ocean basins have undergone enormous and fascinating changes. Let's have a look at some of these.

Pangaea, Laurasia, and Gondwanaland

Scientists believe that the oceans' waters have girdled the globe for about 3 billion years. The distribution of these waters on the surface of the earth has, however, undergone major changes over the eons, as moving land masses have altered the shape of ocean basins.

Slightly over 200 million years ago the terrestrial areas of earth were united in a single supercontinent called *Pangaea* (Figure 4.1) and surrounded by the global ocean, *Panthallassa*, the ancestor of the present-day Pacific. A small gulf called the *Tethys Sea*, which later gave rise to part of the Mediterranean, partially bisected Pangaea at the equator.

About 200 million years ago Pangaea began to break apart, for reasons not yet understood. Fragments of the supercontinent then began to drift away from each other, separated by large rifts from which new sea-floor material flowed. Within 20 million years (i.e., 180 million years before the present or the end of the Triassic period), a new ocean—the Atlantic—had begun to form, even as the

Tethys Sea began to contract. Pangaea had now become two large land masses—*Gondwanaland* to the south and *Laurasia* to the north. During the same period a developing rift, with its upwelling of new sea floor, separated Australia-Antarctica from South America-Africa. India broke completely away from Gondwanaland and began a lengthy journey to the north.

Sixty-five million years of continental drift coincided with the end of the Jurassic period, 135 million years before the present. At about this time Africa cleaved away from South America, giving birth to the South Atlantic Ocean. The Tethys Sea continued to shrink, and the North Atlantic and Indian Oceans continued to grow.

After 135 million years of drift, marked by the end of the Cretaceous period (65 million years before the present), the Tethys had virtually closed to become an inland sea, the Mediterranean. The modern-day shapes of the continents were now clearly evident. Madagascar had separated from Africa, and the South Atlantic was growing.

The last 65 million years have also seen profound changes. India's northward travels have carried it into the continent of Asia, and the two land masses are now united. Their collision has caused folding and uplifting of continental-shelf sediments, in the process creating the Himalayas. Antarctica and Australia separated, becoming distinct continents. Almost half of today's ocean floor has been created in the last 65 million years, and a considerable portion of the Pacific Ocean floor (currently over 1.5 square kilometers per year) is steadily consumed by the subduction of the Pacific Plate beneath the Americas.

The Ocean Basins Today

Today, the earth's lithosphere is divided into seven major plates and a number of minor plates, all moving in a variety of directions relative to one another (Figure 4.8). Gondwanaland and Laurasia have long since fragmented, and their progeny have scattered over the globe. As continental segments have moved on the surface of the earth, the global ocean, Panthallassa, has given rise to the Pacific and Indian Oceans. A new ocean, the Atlantic, has formed where dry land formerly stood. Even today, active spreading centers associated with diverging plate boundaries create new seas: The relatively young and ever-widening Red Sea, formed by the separation of the Arabian and African Plates at the Carlsberg Ridge, is a case in point (Figures 4.8 and 4.9).

The movement of the earth's plates averages only 1 to 4 centimeters per year and rarely exceeds 15 centimeters per year—an amount barely perceptible over a human lifetime. Despite the ominous rumbles of earthquakes and occasional volcanic outbursts, we are reasonably assured that the familiar features of our global landscape will be around for millennia to come. The geological records, however, reveal an irrefutable conclusion: There is nothing so certain on the earth as change. Today's global landscape would be unrecognizable to an observer from 200 million years ago. We can be sure that the changes of the future will be equally profound.

SUGGESTED READINGS

BALLARD, R. D. 1975. Dive into the great rift. *National Geographic* 147:604-15.

BULLARD, E. 1969. The origin of the oceans. *Scientific American* 221:66-75.

DIETZ, R. S. 1977. San Andreas: an oceanic fault that came ashore. *Sea Frontiers* 23:258-66.

DIETZ, R. S., AND HOLDEN, J. C. 1970. The breakup of Pangaea. *Scientific American* 223:30-41.

GLEN, W. 1975. *Continental drift and plate tectonics.* Columbus, Ohio: Merrill.

HALLAM, A. 1975. Alfred Wegener and continental drift. *Scientific American* 232:88-97.

HEIRTZLER, J. R. 1968. Sea-floor spreading. *Scientific American* 219:60-70.

HURLEY, P. M. 1968. The confirmation of continental drift. *Scientific American* 218:52-64.

MACLEISH, W. H., ed. 1979. Ocean/continent boundaries. *Oceanus* 22(3). (A collection of articles.)

MENARD, H. W. 1969. The deep-ocean floor. *Scientific American* 221:53-63.

POLLACK, H. N., AND CHAPMAN, D. S. 1977. The flow of heat from the earth's interior. *Scientific American* 237:60-76.

Scientific American. 1976. *Continents adrift and continents aground.* San Francisco: Freeman. (A collection of articles on plate tectonics.)

SCLATER, J. G., AND TAPSCOTT, C. 1979. The history of the Atlantic. *Scientific American* 240:156-75.

Chapter Five

The Chemistry and Origin of Ocean Waters

OF ALL THE PLANETS IN THE SOLAR SYSTEM, OURS IS THE ONLY one with large amounts of liquid water. Nearly 1.4 billion cubic kilometers of water cover the surface of the earth, while our planetary neighbors either lack water completely or possess exceedingly small amounts of it in nonliquid form.

What would our planet be like without liquid water? We have only to consider the dusty, cratered, lifeless surface of our satellite, the moon, to have some sense of the answer. No oceans girdle its barren circumference, and no clouds grace its virtually nonexistent atmosphere. Temperature extremes of hundreds of degrees bake the surface by day and freeze it by night. It is a desolate, forbidding place.

The thin film of water on our planet is essential for the existence and maintenance of life on earth. The oceans' waters modify global climate and prevent the enormous diurnal and seasonal temperature fluctuations characteristic of other celestial bodies in our solar system. Water is the principal component of all living organisms and the most essential compound required for life. Many substances required by living organisms are readily dissolved in water, thereby making them more accessible to cells. In short, without water our earth would not be inhabited by the creatures with which we are familiar—if indeed it were inhabited at all.

In this chapter we will explore the molecular structure of water, which accounts for most of water's unique, life-sustaining properties. We will examine the physical and chemical properties of sea water and lay the foundation for subsequent discussions of atmospheric and oceanic circulation and the influence of the oceans on the earth's climate. Finally, we will investigate the origin of the oceans and atmosphere.

SOME BASIC PRINCIPLES OF CHEMISTRY

We cannot fully appreciate the remarkable and vital properties of water without first having some understanding of molecules and the atomic building blocks of which they are constructed.

Atoms

Atoms are the fundamental units of which all matter is constituted. A single atom consists of several subatomic particles, notably positively charged *protons*, negatively charged *electrons*, and *neutrons*, which lack charge. Neutrons and protons are located in the central portion, or *nucleus,* of an atom, giving the nucleus a net positive charge. Electrons dart around the periphery of the nucleus in an orbital "cloud," held in place by their attraction to the positive nucleus. The number of electrons and protons in a non-ionized atom (see below) is equal, making the atom as a whole neutral.

While the diameter of the electron cloud is 10,000 times that of the nucleus, most of an atom's mass lies in its nucleus, since both neutrons and protons are 2000 times heavier than electrons. The total mass of an atom constitutes its *atomic mass* or *isotopic mass.*

An *element* is a substance which is comprised entirely of atoms of one kind. Of the 108 elements discovered to date, 90 occur naturally, and 18 have been created in nuclear reactors or laboratories. Individual elements are distinguished primarily by the number of protons in their nucleus, which in turn constitutes the element's *atomic number.* Occasionally, different atoms of the same element may have different numbers of neutrons in their nucleus, even though their proton numbers are equal. Such variants of the same element are called *isotopes.*

Chemical Compounds, Molecules, and Bonding

Elements may combine to form *compounds.* The masses of the constituent elements of a specific compound are always present in the same proportions.

A *molecule* is the smallest particle of an element or compound which retains the properties of the element or compound. The atoms of a molecule may be of the same or different elements, but the proportions of the various atoms are always the same in a given molecule. Thus, a water molecule consists of two atoms of hydrogen and one atom of oxygen, while molecular oxygen contains two atoms of oxygen.

The various atoms of a molecule or compound are linked through *chemical bonds,* and the process of bond formation (or the breaking of bonds) is known as a *chemical reaction.* Chemical reactions do not necessarily occur between atoms; only atoms with compatible atomic structures can combine. The compatibility is based on the number and arrangement of electrons, or *electron configuration,* possessed by the individual atoms. Let us examine electron configuration in greater detail.

TABLE 5.1 Atomic Structure of Representative Elements

ATOMIC NO.	ELEMENT	NUMBER OF PROTONS	NUMBER OF NEUTRONS IN MOST COMMON ISOTOPE	NUMBER OF ELECTRONS Total	Shell 1	Shell 2	Shell 3
1	Hydrogen (H)	1	1	1	1		
2	Helium (He)	2	2	2	2		
3	Lithium (Li)	3	4	3	2	1	
4	Beryllium (Be)	4	5	4	2	2	
5	Boron (B)	5	6	5	2	3	
6	Carbon (C)	6	6	6	2	4	
7	Nitrogen (N)	7	7	7	2	5	
8	Oxygen (O)	8	8	8	2	6	
9	Fluorine (F)	9	10	9	2	7	
10	Neon (Ne)	10	10	10	2	8	
11	Sodium (Na)	11	12	11	2	8	1
17	Chlorine (Cl)	17	18	17	2	8	7
18	Argon (A)	18	22	18	2	8	8

Electron Configuration

The simplest atom, hydrogen, contains one proton and a single electron somewhere outside the nucleus. Atoms of elements of higher atomic number contain additional electrons, the total number of which, under normal circumstances, corresponds to the element's atomic number (Table 5.1). The precise position of any electron at a particular instant is impossible to determine; however, it is possible to calculate the most probable location of any electron at a given time. Calculations show that electrons occupy specific orbitals with specific "average" distances from the nucleus. For the sake of convenience, chemists often refer to these orbitals as "shells," even though atomic orbitals are not necessarily spherical. Considering only the first 18 elements of the Periodic Table, the first two electrons of an atom occupy the first shell, the next eight the second shell, and the next eight the third shell (Table 5.1; Figure 5.1). Above atomic number 18, electron configuration still corresponds to "shells," but the arrangement is more complex.

An atom's bonding potential, or *reactivity*, is determined by the number of electrons in its outer shell. If the outer shell is completely full, the atom is quite nonreactive or "stable." Stable elements are exemplified by the *noble gases*, including helium, neon,

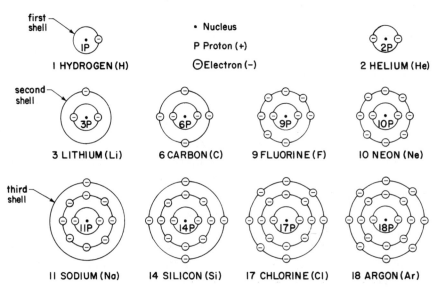

FIGURE 5.1. Electron configurations of atoms of selected elements. Atomic numbers are indicated beside element name. Numbers of protons are shown next to the nucleus of each atom. (After Thurman, H. V. 1981. *Introductory oceanography*. 3rd ed. Columbus, Ohio: Merrill.)

and argon (Figure 5.1). In contrast, atoms with incomplete outer electron shells are more readily capable of bonding, but they are discriminating about their choice of bonding partner. What is the basis for this discrimination? Since a full outer shell represents the most energetically stable configuration for an atom, atoms are most likely to undergo reactions which enable them to achieve this state. Most often this is accomplished in one of two ways: Either the atom donates or receives one or more electrons to or from another atom in a process known as *ionic bonding*, or electrons are shared by atoms in *convalent bonding*. Let's examine some specific examples of each of these processes.

IONIC BONDING. Perhaps the most familiar ionically bonded compound is common table salt, sodium chloride (NaCl). The sodium atom (Na) contains eleven electrons, but only one electron occupies its third, outer shell. Chlorine has seventeen electrons, with seven in its third, outer shell. To achieve a stable configuration, it is obviously to sodium's "advantage" to lose one electron, while chlorine would benefit by gaining one. Is it surprising then that, in mutual proximity, sodium and chlorine are highly reactive (Figure 5.2)?

How does the bonding between the two atoms actually occur? The donation and acceptance of an electron by sodium and chlorine, respectively, mean that the former atom is left with a single net positive charge (eleven protons and ten electrons), while the latter achieves a single net negative charge (seventeen protons and eighteen electrons). The net charges (+1 for sodium and −1 for chlorine) are thus opposite and equal, making the atoms mutually attractive. The resultant bond is an *ionic bond.* In general, the fewer the number of electrons an atom must give up or receive to achieve a stable configuration, the greater its reactivity, since it is energetically more difficult to "trade" three or four electrons than one or two.

Under certain circumstances, ionically bonded compounds may separate into constituent charged particles, called *ions.* For example, sodium chloride may dissolve in water to form the positively charged ion (or *cation*), Na^+, and the negatively charged ion (or *anion*), Cl^-.

COVALENT BONDING. In attaining a completely filled outer electron shell, some atoms are neither as generous nor as demanding as those taking part in an ionic bond arrangement. These atoms find it more convenient to *share* electrons, to the mutual benefit of each participant. Such an arrangement constitutes a *covalent bond.* An example is the bonding arrangement of atoms in a water molecule (Figure 5.3). How does it work? Consider first the hydrogen atoms. Each contains only a single electron. To achieve a full (first) outer shell, each hydrogen atom needs one electron. The oxygen atom requires two additional electrons to completely fill its second, outer shell, which can hold eight. The hydrogen atoms conveniently supply these "missing" electrons. The hydrogens, in turn, can use two of the electrons in oxygen's outer shell (one electron for each hy-

FIGURE 5.2. Ionic bonding in a sodium chloride molecule.

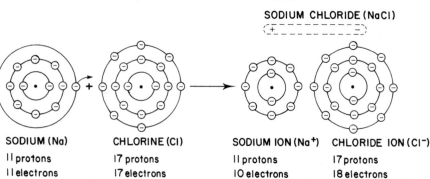

SODIUM CHLORIDE (NaCl)

SODIUM (Na)	CHLORINE (Cl)	SODIUM ION (Na⁺)	CHLORIDE ION (Cl⁻)
11 protons	17 protons	11 protons	17 protons
11 electrons	17 electrons	10 electrons	18 electrons

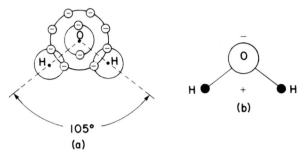

FIGURE 5.3. Covalent bonding and structure of the water molecule.

drogen); thus, by sharing two pairs of electrons, two hydrogen atoms can effectively bond to an oxygen atom.

THE CHEMISTRY AND PROPERTIES OF WATER

The brief introduction to the water molecule brings us to a discussion of a third kind of bond which is important to water and accounts for that substance's many unusual properties. This is the *hydrogen bond*.

Hydrogen Bonding in Water Molecules

The bonding of individual atoms in a water molecule results in an unusual configuration. Instead of being on exactly opposite sides of the larger oxygen atom, separated by 180°, the hydrogen atoms are positioned toward one side, 105° apart (Figure 5.3). Partly because of this arrangement there is a greater concentration of electrons around the oxygen side of the molecule than around the hydrogen side. This gives the oxygen side a slightly negative charge, while the hydrogen side, with its somewhat exposed protons, is slightly positive. Thus, a *polarity* of electrical charge (oppositely charged sides with a space between) is set up within the water molecule; the compound can accurately be called a *polar molecule*.

Because water molecules are polar, their partially charged sides are attracted to oppositely charged sides of adjacent water molecules. The resultant bond is known as a hydrogen bond. While not as strong as ionic or covalent bonds, hydrogen bonds enable water molecules to form the intricate networks characteristic of solid and liquid water, and require significant amounts of energy to break.

Physical Properties of Water

With this background on the molecular structure of water, we can now examine the physical properties of water in greater detail. The unusual thermal properties of water are especially significant.

THERMAL PROPERTIES OF WATER. Water exists in solid, liquid, and gaseous states on our planet, but it is the liquid state which is by far the most prevalent. We take our copious supply of liquid water for granted, but, in fact, it is something of an oddity. Why? Ordinarily, the lighter a substance the more volatile it is (i.e., the more its molecules move about and the more readily it melts or vaporizes). Thus, light materials should have relatively low melting and boiling temperatures. In general, this is true. Water, however, does not fit the pattern. It has a melting and boiling point nearly 100° C higher than methane, which has a similar molecular weight, and significantly higher than a number of substances with much higher molecular weights. Thus, unlike water, most of these substances exist only in gaseous state at the earth's prevailing temperatures and pressure.

Similarly, water has an unusually high *heat capacity*. The heat capacity of a substance is a measure of the amount of heat which must be removed to cool the substance by some measurable amount. Heat capacity is characteristically described in units of heat energy called *calories*. A calorie is defined as the amount of heat energy required to raise the temperature of one gram of liquid water by 1° C. The heat capacity of a gram of liquid water is, therefore, one calorie per degree Celsius, higher than that of any other naturally occurring liquid or solid on earth, except ammonia. Water's high heat capacity accounts for the fact that bodies of water cool down and heat up slowly compared with the adjacent atmosphere and land masses. As a consequence the oceans act as a thermal governor, moderating maritime climates and minimizing temperature extremes over the entire earth. The heat budget of the ocean system and the influence of the oceans on global climate is discussed in Chapter 6.

Why should water have such unique thermal properties? The explanation lies in its hydrogen-bonding structure. A great deal of the heat energy absorbed by water goes into breaking hydrogen bonds and is not manifested as a temperature increase. To under-

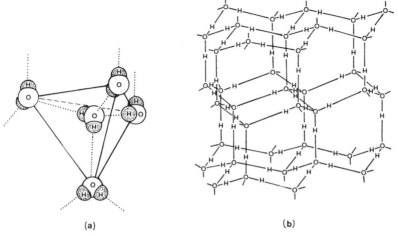

FIGURE 5.4. (a) Tetrahedral arrangement of water molecules in ice. Dotted lines indicate hydrogen bonding between adjacent molecules. (From Brady, J. E. and Humiston, G. E. 1978. *General chemistry: principles and structure*. 2nd ed. N.Y.: Wiley.) (b) The crystalline structure of ice established by the tetrahedral arrangement of hydrogen bonding between water molecules. When ice melts, much of the structure is retained in liquid water. (From Diehl, H. 1974. *Quantitative analysis*. 2nd ed. Ames, Iowa: Oakland Street Science Press.)

stand this, consider what happens when we heat a one-gram block of ice, previously supercooled to −40° C. The application of sufficient heat energy to any substance, including water, will result in a measurable temperature increase (*sensible heat*) of the substance because of the increased kinetic energy (energy of motion) of individual molecules. As we apply heat to our block of ice, its temperature increases proportionately (by 2° C for every calorie of added heat energy, since the heat capacity of *solid* water is 0.5 calorie per gram per degree Celsius), but far less dramatically than for frozen forms of other naturally occurring substances on earth. This is because the presence of hydrogen bonds in solid or liquid water restricts molecular movement.

When our frozen one-gram block of ice warms to 0° C, a puzzling thing happens. Despite considerable additions of heat energy the temperature fails to increase above 0° C. Eighty more calories must be applied and all the ice must be melted before the temperature begins to rise again. The explanation for this delayed response is again related to hydrogen bonding. The "extra" 80 calories of heat energy have been used to break a sufficient number of hydrogen bonds to convert water from the solid to the liquid state. Water molecules in solid ice form a three-dimensional crystalline structure in which the water molecules are equidistantly spaced, through hydrogen bonding, from each other (Figure 5.4). Only when sufficient numbers of these bonds are broken is ice converted into a liquid state which can again undergo temperature increase in response to further applications of heat energy. The "extra" heat needed to break these hydrogen bonds to effect a change of state from solid to liquid water is known as *heat of melting.* An equivalent amount of heat (*heat of fusion*) will be released to the environment when water undergoes the reverse change of state from liquid to solid. Note that there is substantial hydrogen bonding in liquid water, but the bonding is of a weaker, less directional nature than that in ice.

When our now-melted block of ice undergoes a change of state from liquid to gas at a temperature of 100° C, an even greater amount of heat energy (540 calories) is absorbed without an increase in temperature, since *all* hydrogen bonds must be broken to completely vaporize water. This additional heat energy constitutes the *heat of vaporization* of water, which may also be measured as the amount of heat released when a gram of water vapor condenses. Note that water does not need to be heated to the boiling point to effect a phase change from liquid to gas. If sufficient energy is applied, water can vaporize even at 0° C. The vaporization of water at temperatures below the boiling point is known as *evaporation.* Most evaporation of water on earth takes place at the prevailing tempera-

FIGURE 5.5. Phase and temperature changes of water as a function of heat energy. The heat capacity of each phase is given in calories/gram/degree. Note that most heat energy is required to effect phase changes. (Adapted from Gross, G. M. 1982. *Oceanography.* 2nd ed. Englewood Cliffs, N.J.: Prentice-Hall.)

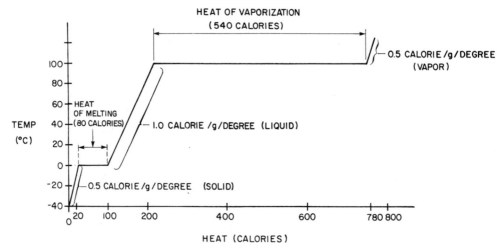

tures on earth (15 to 20° C); at 20° C the *heat of evaporation* (the amount of heat energy required to vaporize water at a temperature below 100° C) is 585 calories.

Let's recapitulate the process of converting our block of ice to water vapor, depicted in Figure 5.5. A total of 740 calories was required to transform one gram of ice, cooled to −40° C, to an equal mass of water vapor at 100° C. Four-fifths of the total, 620 calories, was needed to effect phase changes without concurrent changes in temperature. The heat energy involved in these changes of state was used to break hydrogen bonds between water molecules.

Several other unusual properties of water are attributable to the polarity of its molecules and their resultant hydrogen bonding. These include water's high surface tension, and solvent properties.

SURFACE TENSION. Have you ever watched a water bug stride gracefully across the surface of a placid pond and wondered why it did not break through? In fact, what you witnessed was an example of water's unusually high surface tension—the highest of any common liquid except mercury. The elevated surface tension is attributable to the strong attraction of water molecules to one another at the surface of the water and accounts for several other unusual, but easily observable, phenomena. The next time you pour yourself a glass of water, carefully fill it, drop by drop, as high as possible without overflowing. You will see that the water will actually accumulate above the vessel's rim, forming a convex surface in the process. Very gently lay a needle on the water's surface. If you are careful, the needle will "float," despite the fact that steel is five times denser than water.

The hydrogen bonds of water are responsible for its high surface tension. Their presence means that water molecules at the upper surface are not as strongly attracted to the overlying air molecules as to each other and to the water molecules below them. In addition, air has a relatively low molecular density. Consequently, water molecules clump together and form a slight mound at the air-water interface. The tension which results from the cohesion or tight packing of water molecules at the water surface essentially creates a surface film which makes it more difficult for objects to break through.

SOLVENT PROPERTIES OF WATER. The polar nature of water molecules is also responsible for the fact that water is an extremely good

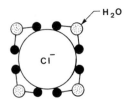

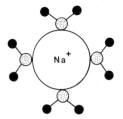

FIGURE 5.6. The hydration of sodium and chloride ions in solution. Each ion is attracted to the oppositely charged poles of water molecules.

solvent. A solvent is a substance which can dissolve other substances or cause their individual atoms, molecules, or ions to separate. The substances which are dissolved are called *solutes*. The combination of solute and solvent is termed a *solution*.

Many polar covalent molecules and many ionic compounds (salts) dissolve readily in water because their charged portions tend to be more strongly attracted to the oppositely charged end of a water molecule than to the remainder of their own molecules. Water literally pulls these molecules or ionic materials apart. They are consequently known as *hydrophilic* ("water-loving"). When salt is poured into water, the polar ends of water molecules separate the sodium and chloride ions and surround them, in a process known as *hydration* (Figure 5.6). Nonpolar molecules, including fats and oils, resist bonding to water molecules and do not go into solution. Such molecules are called *hydrophobic* ("water-fearing").

Water has only a limited capacity to dissolve many substances. When the maximum mass of solute is dissolved in the water, the water is said to be *saturated* with solute. If the level or concentration of solute required for saturation is exceeded, the excess solutes will *precipitate* out of solution, accumulating in the bottom of the container (Figure 5.7). Try this yourself sometime by gradually stirring salt into water and noting the point at which the salt no longer completely dissolves. The *solubility* (capacity for being dissolved) of salts is increased by temperature and pressure.

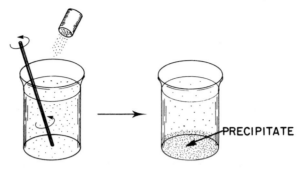

FIGURE 5.7. Precipitation of solutes from a saturated solution.

THE CHEMICAL CONSTITUENTS OF SEA WATER

Sea water is by far the most abundant form of water on earth. It is a solution of virtually every substance on the globe, containing at least 80 of the 90 known naturally occurring elements. Dissolved constituents of sea water include a variety of relatively abundant solids (principally the component ions of salts), gases, "nutrients," and trace elements. The relative proportions of these components are displayed in Figure 5.8. Let's first examine the major dissolved solids which bring about the saltiness or *salinity* of sea water.

Salinity

Salinity may simply be defined as the total weight of dissolved solids (principally salts) in a given weight of water. (Salinity is formally defined as the total amount of dissolved solids in sea water in

FIGURE 5.8. Approximate relative proportions of the dissolved constituents of sea water. (From Stowe, K. 1983. *Ocean science*. 2nd ed. N.Y.: Wiley.)

parts per thousand by weight when all the carbonate has been converted to oxide, the bromide and iodide to chloride, and all organic matter is completely oxidized.) Salinity is commonly measured in grams per kilogram or parts per thousand (0/00 or ppt).

Two aspects of salinity merit special mention. First, of the numerous components of salinity only six comprise over 99 percent of the total. These are chloride (approximately 55 percent), sodium (approximately 30.6 percent), sulfate (approximately 7.7 percent), magnesium (approximately 3.7 percent), calcium (approximately 1.2 percent), and potassium (approximately 1.1 percent). Second, while the saltiness of the oceans is highly variable in time and space, the relative proportions of the various components are remarkably uniform. Thus chloride, the most abundant ion in sea water, will always constitute 55.04 percent of the weight of the dissolved solids in any sea-water sample.

As a consequence, it is possible to measure the concentration of only one constituent of sea water in order to deduce total salinity. Since chloride is the most abundant constituent of salinity, salinity could be calculated by multiplying the chloride content by 1.817 (100/55.04). In fact, salinity is precisely calculated by measuring the total halogen content of sea water (this includes chloride, fluoride, bromide, and iodide) and multiplying this by 1.80655. The total halogen content is referred to as the *chlorinity*. The average chlorinity of the oceans is 19.2 parts per thousand; thus, the seas' average salinity is 34.7 parts per thousand (19.2 times 1.80655).

Since chlorinity determinations of salinity are time-consuming and difficult to carry out at sea, salinity measurements are often conducted using electronic instruments. A *salinometer* calculates salinity by measuring the electrical conductivity of a sea-water sample. Since conductivity (a function of the concentration of dissolved ions) is directly proportional to the total weight of dissolved solids, a conversion to salinity is easily accomplished. Calibration is routinely carried out using ampules of standard sea-water samples, accurate to the nearest one part in ten million, which are provided by a British laboratory.

The presence of dissolved solids in water affects ice formation. The addition of salts lowers the temperature at which the water will first freeze (i.e., causes a *freezing point depression*). Not until the temperature of average oceanic sea water (salinity of 34.7 parts per thousand) declines to nearly $-2°$ C does freezing occur (Figure 5.9).

The presence of salts also affects the *density* (mass per unit volume) of water. While a detailed treatment of sea water density is presented in Chapter 7, some discussion is pertinent here.

In general, water density increases with declining temperature and increasing salinity and/or pressure. Surprisingly, however, pure, fresh water at atmospheric pressure achieves its maximum density at 4° C. Below that temperature, the incipient formation of ice crystals and the resultant lattice work arrangement of hydrogen bonds (Figure 5.4) spreads the water molecules a little further apart than they were at 4° C, thus slightly decreasing water's density. This decrease in density continues until the water turns completely to ice. Therefore, ice floats on water despite its colder temperature. This is fortunate. Were the temperature of maximum density to coincide with the freezing point of fresh water, ice would form from the bottom up in lakes and ponds (since the densest water would sink to the bottom) and would take much longer to thaw with warming air temperature. Furthermore, the surface layer of ice on freshwater bodies serves to insulate the warmer, denser liquid water below to the advantage of bottom-dwelling residents.

The addition of salts lowers water's temperature of maximum density (Figure 5.9) until, at a salinity of 24.7 parts per thousand, it coincides with the freezing point (at −1.33° C).

As sea water cools, its increased density causes it to sink. When surface waters have cooled to their freezing point, the formation of ice crystals excludes about 70 percent of the salt from sea ice. The salt accumulates below the ice, increasing the density of the non-

FIGURE 5.9. The effect of dissolved solids on the initial freezing point and the temperature of maximum density of water. (Adapted from Anikouchine, W. A., and Sternberg, R. W., 1973. *The world ocean.* Englewood Cliffs, N.J.: Prentice-Hall.)

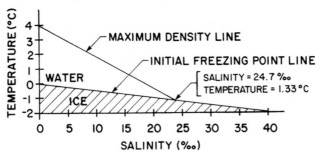

TABLE 5.2 Selected Factors Affecting the Concentration of Gases in Sea Water

GAS	FACTOR	EFFECT
	Physical-Chemical Factors	
Any	Wind, waves, turbulence	Enhances exchange with atmosphere.
Any	Concentration differences between atmosphere and sea	Gas flux follows concentration gradient until equilibrium reached.
Any	Temperature	Decreasing temperature increases solubility.
Any	Pressure	Increasing pressure increases solubility.
Any	Salinity	Increasing salinity decreases solubility.
Any	Interactions with other dissolved compounds (e.g., $CaCO_3$)	Depends on compound and conditions (see text).
CO_2	Buffering capacity of sea water	Ocean absorbs CO_2 and buffers (see text).
CO_2	pH of sea water	Affects "form" of carbon dioxide in sea water (see text).
	Biological Factors	
CO_2	Decomposition	Increases CO_2 concentration.
CO_2	Respiration	Increases CO_2 concentration.
CO_2	Photosynthesis	Decreases CO_2 concentration.
O_2	Decomposition	Decreases O_2 concentration.
O_2	Respiration	Decreases O_2 concentration.
O_2	Photosynthesis	Increases O_2 concentration.

frozen water and increasing its sinking rate. The low-salinity (and low-density) ice which forms and floats at the water's surface insulates the waters below and slows down the freezing process. This prevents the polar seas from freezing solid.

Dissolved solids constitute the overwhelming fraction of substances dissolved in sea water, but there are a number of other constituents which, though of miniscule concentrations, are essential for the sustenance of life in the oceans. Among the most important of these are *dissolved gases*.

Dissolved Gases

By and large, the gases in sea water are the same as those in the atmosphere but their concentrations and relative proportions are very different. For example, nitrogen (as N_2) is by far the most abundant gas in the atmosphere, constituting 78 percent of the total atmospheric gases, 2600 times the concentration of CO_2. Nitrogen gas is also abundant in the oceans (comprising 47.5 percent of the gases in surface seawater), but the proportion of CO_2 in the oceans (15.1 percent of the surface sea water gases) is much greater than in the atmosphere.

The atmosphere is the ultimate source of gases in sea water, and exchanges at the air-water interface are a function of physical and chemical interactions between the two environments (Table 5.2). Once the gas-containing sea water leaves the surface, its gas concentrations may be affected by biological processes as well as physical and chemical ones (Table 5.2).

The *saturation value* for a gas in solution refers to the maximum amount of that gas which can be retained in solution without loss to the atmosphere. When saturation is achieved, the gas in question is present in *equilibrium concentration*. Saturation values increase with increasing water pressure and decreasing temperature and salinity, with temperature usually having the most influence in the oceans.

Gases are incorporated into sea water from the atmosphere through two principal (and universal) physical mechanisms. The first process is *diffusion*, the random movement of molecules from a region of higher concentration to a region of lower concentration (Figure 5.10). While the movement of individual molecules is quite rapid (and is primarily a function of temperature), the *net* movement of molecules is slow because the random nature of the movement means that an individual molecule has an equal probability of

FIGURE 5.10. The diffusion of solutes in a liquid. Given enough time, concentrated particles will become uniformly distributed through diffusion alone. Stirring will accelerate the process. See text for fuller explanation.

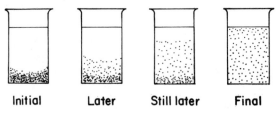

Initial   Later   Still later   Final

going in any conceivable direction at any time. The diffusion rate of molecules in water is only 1/10,000 that of the rate in air.

The other process of incorporating gases into water is through physical mixing from the atmosphere. The mixing is accomplished primarily by winds, waves, and currents.

The most biologically important gases in sea water are carbon dioxide (CO_2) and oxygen (O_2).

CARBON DIOXIDE. Once CO_2 diffuses or is mixed into sea water, it undergoes a number of reactions with water molecules and other dissolved compounds as follows:

$$H_2O + CO_2 \leftrightarrows H_2CO_3 \leftrightarrows H^+ + HCO_3^- \leftrightarrows H^+ CO_3^{2-} \tag{1.1}$$

where H_2CO_3 is carbonic acid, HCO_3^- is bicarbonate ion, and CO_3^{2-} is carbonate ion. Since these reactions are in large part dependent on the *acidity*, or *pH*, of seawater, let's have a closer look at pH.

Substances which, when dissolved in water, release free hydrogen ions and an anion into the medium are known as *acids*. A strong acid readily dissociates in water, and a weak acid dissolves more reluctantly. An example of a strong acid is hydrochloric acid (HCl), which, in water, breaks down into a free hydrogen ion (H^+) and the negatively charged chloride ion (Cl^-). A weak acid is exemplified by carbonic acid (H_2CO_3).

The acidity of a solution is a measure of the amount of hydrogen ions (H^+) in the solution. The more free H^+, the more acidic the medium. Since it is impossible to count all H^+ ions, their concentration is determined instead by calculating the number in a known quantity of water. For example, one hydrogen ion in 10,000,000 molecules of water would result in a H^+ ion concentration of 1/10,000,000 or .0000001 or 10^{-7}. More simply, we can state this as follows: $[H^+] = 10^{-7}$. The brackets around H^+ should be read as "the concentration of H^+."

In water a certain number of the individual water molecules undergo reactions with their constituents as follows:

$$H_2O \leftrightarrows H^+ + OH^- \tag{1.2}$$

The OH^- ion is known as a *hydroxide* ion. Since, in pure water, these reactions are in equilibrium, every H^+ ion is balanced by a

TABLE 5.3 The Relationship of pH, Hydrogen Ion Concentration ($[H^+]$), and Hydroxide Ion Concentration ($[OH^-]$) in Solutions of Varying Acidities.

pH	$[H^+]$	RELATIVE ACIDITY/ ALKALINITY	$[OH^-]$	PRODUCT
1	10^{-1}	Acid	10^{-13}	10^{-14}
2	10^{-2}	Acid	10^{-12}	10^{-14}
3	10^{-3}	Acid	10^{-11}	10^{-14}
4	10^{-4}	Acid	10^{-10}	10^{-14}
5	10^{-5}	Acid	10^{-9}	10^{-14}
6	10^{-6}	Acid	10^{-8}	10^{-14}
7	10^{-7}	Neutral	10^{-7}	10^{-14}
8	10^{-8}	Alkaline	10^{-6}	10^{-14}
9	10^{-9}	Alkaline	10^{-5}	10^{-14}
10	10^{-10}	Alkaline	10^{-4}	10^{-14}
11	10^{-11}	Alkaline	10^{-3}	10^{-14}
12	10^{-12}	Alkaline	10^{-2}	10^{-14}
13	10^{-13}	Alkaline	10^{-1}	10^{-14}
14	10^{-14}	Alkaline	10^{0}	10^{-14}

hydroxide ion, and the solution is said to be neutral. In pure water the concentration of H^+ ions is 10^{-7}; thus, the concentration of OH^- ions is also 10^{-7}. The pH of a solution is defined as the negative logarithm (*log*) of the H^+ ion concentration; consequently, since the log of 10^{-7} is -7, the pH of pure water is 7.

The product of $[H^+]$ and $[OH^-]$ in any solution at 25° C is always 10^{-14}. Thus, in a solution in which $[H^+]$ equals 10^{-3}, there would be one H^+ ion for every 1000 (10^3) water molecules and one OH^- ion for every 10^{11} water molecules. The pH of the solution would be 3. The relationship of $[H^+]$, $[OH^-]$, and pH is shown in Table 5.3. Please note that since the pH scale is logarithmic, every unit change in pH reflects a tenfold $[H^+]$ change. Thus, a pH of 6 is ten times as acid as a pH of 7; a pH of 5 is ten times as acid as a pH of 6 and one hundred times as acid as a pH of 7. Note also that at a pH of less than 7 a solution is said to be *acidic*, while at a pH greater than 7 it is basic or *alkaline*. The pH scale ranges between 0 and 14.

With this background, let's have another look at the behavior of carbon dioxide in sea water. Recall the equilibrium reactions and various forms of carbon dioxide in sea water described above. The pH of sea water is normally slightly alkaline (7.8 to 8.1) because of the presence of HCO_3^-, CO_3^{2-}, and other basic anions. Up to a

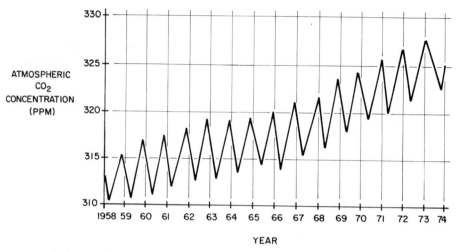

FIGURE 5.11. Changes in global atmospheric carbon dioxide concentration: 1958–1974. (From Calder, N. 1974. *The weather machine*. N.Y.: Viking.)

point, the introduction of additional CO_2 to sea water, through biological or physical processes, simply results in greater production of H_2CO_3, with little or no effect on pH. Similarly, the removal of CO_2 (through photosynthesis, for example) results in a compensatory production of more H_2CO_3 by shifting the reaction in (1.1) to the left from carbonate and bicarbonate ion to carbonic acid production. Again, pH is maintained within relatively narrow limits.

Carbonic acid, in combination with bicarbonate and carbonate, also acts as a *buffer* against the introduction of acid or base to sea water. When more H^+ is introduced and pH is temporarily lowered, the equilibrium reactions shift to the left; more hydrogen ions are incorporated into bicarbonate ions and carbonic acid, producing a corresponding increase in pH (or decrease in acidity). If more OH^- is introduced, the reaction will shift in the other direction, consuming H^+ to react with the added base and minimizing changes in pH. It should be noted that this buffering capacity, while significant, is not unlimited; at some point sea water can no longer compensate for the addition of excess acids or bases.

The buffering capacity of sea water has had profound implications for the fate of atmospheric CO_2, particularly since the advent of the industrial age. The burning of fossil fuels has released enormous quantities of CO_2 since the 1850s, with most of this occurring during the past few decades (Figure 5.11). It is estimated that the oceans have absorbed about half this quantity apparently without ill effects. However, the oceans' absorptive capacity is limited. Consequently, there is now acute concern that significantly increased atmospheric CO_2 accumulations may lead to a marked increase in the temperature of the earth (with a possible catastrophic rise in sea level as polar ice and glaciers melt) because of the so-called *greenhouse effect*. This topic will be discussed in greater detail in Chapter 6.

Sea water–carbon dioxide chemistry is also affected by complex interactions between carbon dioxide and calcium which are largely beyond the scope of this book. The oceans' waters are, for the most part, saturated with calcium ion (Ca^{2+}). Calcium is a component of calcium carbonate ($CaCO_3$), which makes up the shells and skeletons of many marine organisms. In warm, shallow waters, reduced CO_2 solubility may result in the formation of a solid product of the equilibrium reaction of free calcium and carbonate ions in sea water as follows:

$$Ca^{2+} + CO_3^{2-} \leftrightarrows CaCO_3 \tag{1.3}$$

When solid $CaCO_3$ becomes sufficiently abundant, it precipitates out of sea water, forming deposits on the bottom which may eventually become limestone. This occurs over wide portions of the Bahamas Bank, the Great Barrier Reef, and the Persian Gulf.

Conversely, in darker, colder ocean waters, CO_2 concentrations are relatively high (because of decomposition and respiration of dead and living organic matter, the absence of CO_2-dependent photosynthesis, and increased solubilities because of low temperature and high pressure). These conditions favor dissolution of $CaCO_3$. In general, complete dissolution of $CaCO_3$ occurs at depths below about 4500 meters. This depth is commonly referred to as the *calcium carbonate compensation depth*.

OXYGEN. Oxygen concentrations in sea water are affected by exchanges with the atmosphere and by biological processes (Table 5.3). The addition of oxygen can result from diffusion or mixing from the atmosphere or from photosynthesis, the process whereby green plants make energy-rich organic compounds from inorganic constituents using the sun's energy (see Chapter 9). Oxygen gas is a byproduct of photosynthetic reactions and is released into the surrounding air or water.

Because photosynthesis-stimulating light only penetrates through the top 100 meters or so of ocean, and because most of the oceans' depths are too far removed from the atmosphere for direct physical exchange of gases, oxygen can be added to the oceans only in the surface waters. However, oxygen is consumed at all depths through the biological processes of decomposition and respiration, which oxidize organic matter (see Chapter 9). Because the abundance of living organisms decreases with depth, the relative importance of decomposition in oxygen consumption increases with depth.

These processes result in a steady decline in oxygen concentration from surface waters to a depth of 150 to 1000 meters. Below this depth (the *oxygen minimum zone*) oxygen concentrations gradually increase with depth because low water temperatures mean high gas solubility, because oxygen-demanding living and dead organic matter declines with depth, and because deep waters originally sank from polar surface waters where they were saturated with oxygen (see Chapter 7).

In some bottom waters with restricted circulation (e.g., deep ocean basins, fjords, and stagnant estuaries), oxygen concentrations may become completely exhausted, resulting in *anaerobic* (lacking oxygen) conditions.

Other Dissolved Constituents of Sea Water

In addition to the major constituents of salinity and the important dissolved gases, many minor substances are dissolved in sea water in vanishingly small amounts. While their concentrations are measured only in parts per million or even parts per billion, some are important enough to have a vital influence on the populations and distributions of organisms in the ocean. Since the concentrations of biologically important substances may fluctuate considerably in time and space, they are said to be *nonconservative properties* (in contrast to the "conservative" nature of the major salts of the sea, whose concentrations are seldom significantly affected by biological processes). The minor constituents of sea water include *nutrients*, *trace elements*, and *dissolved organic matter*.

NUTRIENTS. Nutrients consist of dissolved inorganic substances which are essential for the growth and development of all photosynthetic organisms, including those in the sea. Generally, these substances exist in sea water in concentrations of only parts per million or fractions thereof. The nutrients which are in most demand by marine plants (and most likely to be scarce enough to limit plant growth) are ions of nitrogen (available primarily in the form of dissolved ammonium, NH_4^+, or nitrate, NO_3^-) and phosphorus (as phosphate, PO_4^{3-}). In addition, *diatoms*, an important group of single-celled algae (see Chapter 9) require significant amounts of silicate ion ($Si(OH)_4^{4-}$) in the construction of their cell walls. The importance, utilization, and cycling of nutrients will be discussed in greater detail in Chapter 9.

TRACE ELEMENTS. Only eleven elements (including the six described above) constitute 99.99 percent of all of the dissolved solids in the sea. The other elements dissolved in sea water usually exist in such minute concentrations (in the parts-per-billion range or fraction thereof) that they are referred to as *trace elements*. The study of these substances is an active field today, since some have important effects on oceanic life. For example, some trace elements are essential for growth and reproduction. Furthermore, marine organisms may biologically accumulate trace elements so that tissue concentrations far exceed those in surrounding sea water and may even be toxic to predators, including humans. The well publicized biological concentration of mercury by swordfish is a case in point.

DISSOLVED ORGANIC MATTER. A host of dissolved organic substances may be found in sea water, including carbohydrates, proteins, lipids, vitamins, and hormones. These and their breakdown products derive from decomposition of animal and plant tissue or represent excreted materials from living organisms. There is growing interest in dissolved organic matter in the sea because of developing awareness of its abundance and its potentially important role in the nutrition of microscopic organisms.

Residence Time of Materials in the Oceans

Earlier, we made the statement that the proportions of various elements remain constant in sea water, even though salinity itself is variable. We also have evidence, based on the chemical composition of fossils, that ocean salinity has remained virtually constant for the past half billion years. Therefore, the composition of the oceans may be described as representing a *steady-state condition*. For this steady

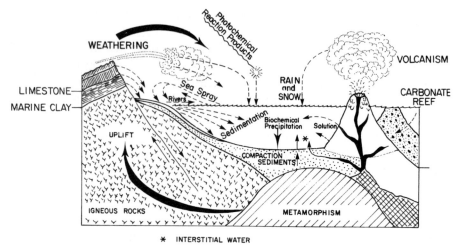

WEATHERING

Photochemical Reaction Products

VOLCANISM

LIMESTONE—
MARINE CLAY—

RAIN and SNOW

CARBONATE REEF

Sea Spray

Rivers

UPLIFT

Sedimentation

Biochemical Precipitation

Solution

COMPACTION SEDIMENTS

IGNEOUS ROCKS

METAMORPHISM

∗ INTERSTITIAL WATER

FIGURE 5.12. Sea-water substances: their origins, fates, and cycling between the oceans, land, and atmosphere. (From Anikouchine, W. A., and Sternberg, R. W. 1973. *The world ocean*. Englewood Cliffs, N.J.: Prentice-Hall.)

state to be maintained, each of the dissolved solids, and water itself, must be replaced in the oceans as rapidly as it is removed. Removal processes include incorporation into living organisms, binding into sediment and suspended particles, evaporation of sea water, and the reactions of sea water with materials erupting from the sea floor. Replacement processes include biological reactions, decomposition of organic materials, weathering of terrestrial crust (primarily introduced to the oceans by rivers), emission from volcanoes and hot springs, reactions with sediments, and precipitation of suspended particles (Figure 5.12).

The *residence time* of a substance is the time that it takes to completely replace or renew that substance in the ocean. Residence time can be calculated by dividing the total amount of that substance by its rate of removal (or rate of replacement). Residence times of oceanic substances are related to their reactivity and range from 100 years for aluminum to 260 million years for sodium. Water has a residence time of about 40,000 years, based on a worldwide evaporation rate of 10 centimeters per year and an average ocean depth of 4000 meters.

THE ORIGIN OF THE OCEANS

In its formative stages, the earth was too hot to have retained the relatively light molecules, including water, which make up today's atmosphere and oceans—these volatile materials would simply have boiled away. Eventually, however, the earth cooled enough for the molten rocks on its surface to solidify (as the earth's crust) and to allow water vapor and gases to be retained on and above the earth's surface. According to most estimates, this occurred about four and a half billion years ago and essentially signifies the true "origin" of the earth.

Once the earth had cooled sufficiently, the atmosphere could hold water vapor and the seas could evolve. Where did the oceans' waters and salts come from, and how quickly did they accumulate? We might first look to the early atmosphere as the source of ocean waters, but that could not have been the direct source. The oceans today hold 1370 million cubic kilometers of water, but the atmosphere contains only a miniscule fraction of that amount. Even if the early atmosphere were saturated with moisture, its complete condensation would have contributed, at most, a few centimeters of water to the earth's surface. We must then look to the earth's rocks for the ultimate source of water.

The crust of the earth contains water, but the quantities are much too small to have accounted for the origin of the oceans. While we have not yet determined the *in situ* composition of mantle material, analysis of meteorite rocks, which are presumed to be of similar composition, suggests a water content of 0.5 percent. By estimating the total mass of the mantle it is relatively easy to calculate that the mantle today contains 16 times as much water as the oceans even now hold. Thus, many scientists believe that the mantle has been the origin of most of the oceans' waters.

How has mantle water made it to the surface of the earth? The most obvious mechanisms are volcanic eruptions and hot springs. This process of *outgassing* from the earth's interior also transports other gases and minerals to the surface of the earth.

There has been considerable debate over how rapidly the oceans' waters accumulated on the earth's surface after the planet cooled down sufficiently to retain water. One can approach this problem by calculating how rapidly new sources of water (i.e., "juvenile water") are added to the oceans today; however, measurements of current rates of outgassing are complicated by the fact that much of the water emanating from volcanoes and hot springs appears to be recycled surface water. Nonetheless, there is general agreement that water of mantle origin continues to accumulate in the oceans even today, though probably at a slower rate than in the early years of the earth's history.

What is the origin of the ocean's salts? One major source of these dissolved solids is the earth's crust, whose materials reach the ocean through weathering and erosion. In fact, when we analyze the mineral composition of the crust, atmosphere, oceans, and ocean sediments today, it is clear that some constituents of sea water (e.g., sodium, magnesium, potassium, and calcium) probably derived from the crust. Others, however (e.g., chlorine, bromine, carbon dioxide, nitrogen, sulfur, and fluorine), are not sufficiently abundant in the crust for continental rocks to have been their origin. Because compounds containing these elements exist primarily in gaseous form in the atmosphere, they would have volatilized with water vapor in the earth's hot early atmosphere. Thus, like water, these *excess volatiles* appear to have been outgassed from the mantle. Analysis of gases from volcanic eruptions has confirmed the presence and appropriate ratios of these substances.

SUGGESTED READINGS

GABIANELLI, V. J. 1970. Water—the fluid of life. *Sea Frontiers* 16:258-270.

GROSS, M. G. 1982. *Oceanography: a view of the earth.* 3rd ed. Chapter 5. Englewood Cliffs, N.J.: Prentice-Hall.

HARVEY, H. W. 1960. *The chemistry and fertility of sea waters.* New York: Cambridge University Press.

MacINTYRE, F. 1970. Why the sea is salt. *Scientific American* 223:104-15.

REVELLE, R. 1963. Water. *Scientific American* 209:93-108.

RILEY, J. P., AND CHESTER, R. 1971. *Introduction to marine chemistry.* New York: Academic Press.

STOWE, K. 1983. *Ocean science.* 2nd ed. Chapters 10 and 11. New York: Wiley.

THURMAN, H. V. 1981. *Introductory oceanography.* 3rd ed. Chapters 2 and 9. Columbus, Ohio: Merrill.

Chapter Six

The Oceans and Climate

THE WINTER OF 1982-1983 WAS MARKED BY A SERIES OF GLOBAL weather catastrophes which appear to have been unprecedented in this century. Vicious storms raked California's coast, bringing torrential rains, major flooding, and mudslides and causing hundreds of millions of dollars of property damage, a score of fatalities, and the displacement of 10,000 homeowners. Simultaneously, equally devastating storms swept Ecuador and northwest Peru. In marked contrast, countries in Africa and on the Asian side of the Pacific Basin were stricken by severe droughts, which resulted in billions of dollars of crop losses and widespread starvation. It is now believed that these far-ranging catastrophes were interrelated and attributable to a major oceanographic disturbance in the Pacific Ocean, the *El Niño/Southern Oscillation,* which had directly disruptive effects on the earth's atmosphere.

The El Niño/Southern Oscillation, described in greater detail later in this chapter, is just one example of the increasingly evident influence of the oceans on global weather and climate. Other examples, all of which underscore the close coupling between the oceans and the atmosphere, include the crucial role of the oceans in global carbon dioxide dynamics (which in turn have major effects on the temperature of the earth), the effects of ocean circulation on the heat balance of the world, and the importance of the tropical seas (through evaporative heat loss and the spawning of hurricanes) as "heat pumps" in global atmospheric circulation.

To understand the intimate relationship between the oceans and atmosphere, we should first consider atmospheric circulation in general and the balance of heat on the globe.

THE DISTRIBUTION AND BALANCE OF HEAT ON THE EARTH

Sources of Heat on Earth

As the sun blazes across 150 million kilometers of space, kindling our earthly fires, it is hard to believe that there are other sources of heat on earth. Yet, there are a few. Chemical reactions in the earth's interior contribute small amounts of heat to the surface. Heat produced by frictional contact between the rotating earth and the atmosphere and heat released by other chemical and biological activities

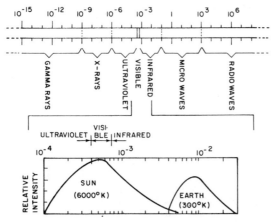

FIGURE 6.1. The spectrum of electromagnetic radiation from the shortest gamma rays to the longest radio waves. Most energy emitted by the sun is concentrated within the visible range (with some in the ultraviolet and near-infrared also), whereas that emitted by the significantly cooler earth lies primarily in the longer infrared range. (Adapted from Stowe, K. 1983. *Ocean science.* 2nd ed. N.Y.: Wiley.)

on earth contribute still less. However, well over 99 percent of the energy absorbed on earth is from the sun. That is the only source we will consider here.

Any body warmer than a temperature of absolute zero (zero degrees Kelvin or −273 degrees Celsius) will radiate energy to its surroundings. The amount of energy radiated is directly related to the temperature of the body but inversely related to the wavelength at which most of the emitted energy is radiated (known as the *wavelength of maximum emission*). The sun, with a temperature of about 6000 ° K, has a wavelength of maximum emission of about 0.5 micron, which falls within the visible light portion of the electromagnetic spectrum. A significant portion of the sun's energy is radiated within the ultraviolet and infrared spectra also (Figure 6.1).

Is the sun's radiant-energy output constant? There is strong evidence that it is not. Astronomers have known for decades that the sun periodically flickers, like a light bulb connected to a variable power source. The flickering, which is largely unpredictable, is attributable to violent solar storms which rage across the sun's surface. These fiery storms, similar to terrestrial hurricanes, are called

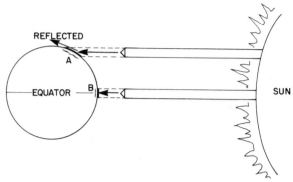

FIGURE 6.2. The sun's energy received on earth varies with latitude in part because an equal amount of heat is spread over an increasingly larger area at increasingly higher latitudes. There is also more reflection of solar energy at higher latitudes, since the sun's rays strike more obliquely there. (After Stowe, K. 1982. *Ocean science.* 2nd ed. N.Y.: Wiley.)

sunspots. They apparently result in fluctuations, of as yet uncertain magnitude, in solar energy received on earth. However, in considering the earth's heat budget, scientists generally assume that the earth as a whole receives a constant energy input from the sun.

The Fate of Incoming Solar Energy

Let's examine what happens to our "constant" input of solar energy from the time of its arrival in the earth's outer atmosphere to the time of its actual absorption on earth. When the sun is directly overhead, the earth's atmosphere receives about two calories of energy per square centimeter per minute (or two "langleys," per minute). This energy flux is often referred to as the *solar constant.* Taking into account that the earth is a rotating sphere which is half in darkness at any one time, and that energy received varies with latitude, the average amount of solar energy reaching the outer atmosphere is about 0.5 langley per minute.

Most of this energy never reaches the surface of the earth. On its way to the ground it is absorbed by ozone, water vapor, clouds, and dust and reflected directly back to space by clouds, dust, air, haze, and the surface of the earth itself. Less than half (47 percent) of incoming solar radiation is actually absorbed, directly or indirectly, by the earth. This absorbed energy is not equally distributed over the globe. In the first place, lower latitudes receive far more energy than higher latitudes over the course of a year. There are several reasons for this. First, the sun's rays strike high latitudes at a comparatively low angle, causing more energy to be reflected than at lower latitudes. Second, a band of solar radiation is spread over a wider area at high latitudes than at low latitudes. Finally, solar energy must pass through a greater volume of atmosphere before it strikes high-latitude environments than before impinging on lower latitudes. These principles are illustrated in Figure 6.2.

The inclination of the earth's axis also contributes to differential heating of the globe over the course of a year because it causes the seasons. The globe rotates at an angle of 23.5 degrees with the plane of the earth's orbit with the sun (the *elliptic;* Figure 6.3). As a consequence areas north of the Arctic Circle and south of the Antarctic Circle receive no direct solar radiation during their respective winter seasons.

The Distribution of Solar Energy on Earth

The unequal distribution of solar energy on the surface of the earth results in an annual net gain of energy between the equator and 35 to 40 degrees north and south, and a net annual loss in higher latitudes. Despite this latitudinal disparity, the temperature of the globe, and of any region on the globe, remains essentially constant from year to year. There are two related reasons for this. First, the earth as a whole generally loses as much heat as it gains during the course of a year. This balance between incoming and outgoing heat establishes the earth's *heat budget.* Second, there is a redistribution

FIGURE 6.3. The tilt of the earth's axis accounts for the changes of seasons as the earth orbits the sun.

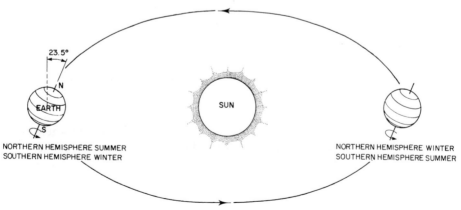

NORTHERN HEMISPHERE SUMMER
SOUTHERN HEMISPHERE WINTER

NORTHERN HEMISPHERE WINTER
SOUTHERN HEMISPHERE SUMMER

of absorbed heat energy on the globe such that equatorial regions do not continuously heat up and polar regions do not continuously cool down. Let's examine each of these phenomena in turn.

THE HEAT BUDGET OF THE EARTH. In its simplest form, a heat budget describes the relationship between incoming and outgoing heat as follows:

Heat Gain − Heat Loss = Net Heat Gain or Loss

The heat budget of the earth states that, over a year, heat gain minus heat loss is zero for the earth in general and for any individual portion of the earth (including the oceans). We have stated that the sun is the only significant source of heat gained. What then are the mechanisms of heat loss?

There are three principal ways in which the earth gives up heat. These are through *back radiation* to the upper atmosphere or space, through *evaporation* of water to the atmosphere, and through *conduction (sensible-heat transfer)* into the atmosphere. Let's examine each of these in turn.

Like the sun, the earth is a body with a temperature above absolute zero, but its average temperature of 290° K (17° C) is far below that of the sun. Therefore, it not only radiates significantly less energy to space than the sun, but its wavelength of maximum emission is much longer—approximately 10 microns, which falls within the infrared range. On average, 41 percent of absorbed solar energy is reradiated by earth to space, either directly or via the earth's atmosphere. This percentage would be higher were it not for the presence of water vapor and carbon dioxide in the atmosphere.

Although the atmosphere is transparent to the short visible wavelengths of the sun, it is relatively opaque to the longer, infrared wavelengths emitted by the earth. Much of this radiation is absorbed by atmospheric water and carbon dioxide, and a significant portion of this "trapped" energy is reradiated to earth. Consequently, the atmosphere acts like an insulating layer around the earth, preferentially retaining some infrared heat and artificially warming the earth's surface in the process. This phenomenon is called the *greenhouse effect* because the principle is the same as that which keeps greenhouses significantly warmer than the air outside. In greenhouses, however, it is the glass which traps outgoing infrared radiation (Figure 6.4). Without a greenhouse effect the earth's

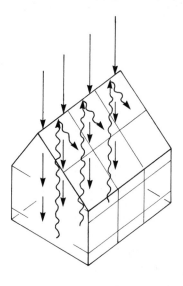

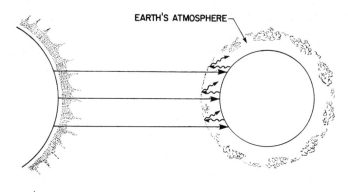

EARTH'S ATMOSPHERE

↓ SOLAR ENERGY (mostly in UV and visible wavelengths)

ᖰ TRAPPED INFRARED RADIATION

FIGURE 6.4. The "greenhouse effect." Like the glass in a greenhouse, the earth's atmosphere is essentially transparent to incoming solar energy but relatively opaque to much of its own outgoing radiation, which is primarily in the infrared range. This keeps the earth much warmer than it would be without an atmosphere.

temperatures might plunge to −20° C, the temperature at the top of the earth's atmosphere. However, there is major current concern that the accelerated burning of fossil fuels is enhancing the greenhouse effect by significantly increasing carbon dioxide concentrations in the atmosphere. We will return to this concern later in the chapter.

The earth also gives up large amounts of heat to the atmosphere through the evaporation of water. Recall the thermal properties of water, described in Chapter 5. One of the most important is water's high heat of vaporization. Let's review this briefly. Compared with other materials, a relatively large amount of heat must be added to water to be reflected in a temperature increase because of the presence of intermolecular hydrogen bonds which restrict the kinetic energy of water molecules. An even greater amount of heat must be applied to cause a change of state of water. For example, it requires 100 calories to raise water temperature from the freezing point to the boiling point but 540 calories to change water from a liquid to a gas at 100° C. The "extra" heat is required to break all hydrogen bonds and is called the *heat of vaporization*. Water can be vaporized or evaporated at temperatures below the boiling point, but it requires even more heat energy to do so. At whatever temperature water is vaporized, the evaporated molecules break free of the liquid water surface and enter the atmosphere as water vapor. Those that break free have absorbed more heat energy than their neighbors left behind; consequently, the remaining liquid water is made cooler. That is why the act of perspiring (which is essentially the evaporation of water from skin) is a cooling process.

Therefore, evaporation removes heat from reservoirs of water on the earth's surface (e.g., oceans, lakes, and leaves of trees) and adds it to the atmosphere. However, the air is not directly heated in the process. The heat contained in water vapor exists as stored or *latent heat*, as opposed to *sensible heat*, which is manifested by a direct atmospheric temperature increase. Latent heat is not released to the atmosphere until water vapor condenses again into liquid water or rain. By this time water vapor has ascended high enough into the atmosphere that the released heat is largely lost to space by radiation and not returned to the earth. Consequently, the evaporation of water is viewed as a one-way process in which heat is transferred from the earth's surface to the atmosphere but not returned. Approximately 53 percent of the heat loss by the earth is via the process of evaporation. The overwhelming proportion of this evaporation takes place in the oceans.

The last important mechanism by which the earth gives up absorbed heat is through *conduction* (sensible-heat loss). Conduction is the process by which heat (thermal energy) is transferred, by intermolecular contact, from a warmer to a cooler contiguous surface. Most of the heat energy lost by the earth through conduction is by transfer from the oceans to the atmosphere. As with evaporation, this transfer is mostly a one-way process with conducted

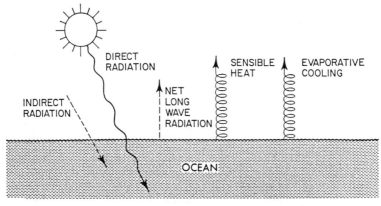

FIGURE 6.5. Major components of the earth's heat balance. (From Bryan, K. 1978. The ocean heat balance. *Oceanus* 21(4):20. Reprinted by permission of the Woods Hole Oceanographic Institution.)

heat ultimately radiated from the atmosphere to space. Even when the air is warmer than the ocean (which it frequently is), the tendency of heat to rise means that a cool, insulating layer of atmospheric air will sink to just above the water's surface, effectively blocking the transfer of heat from warmer air above. Conduction transmits far less heat loss to the atmosphere (and, ultimately, to space) than back radiation and evaporation, representing only about 6 percent of the total.

A schematic diagram of heat balance at the oceans' surface is presented in Figure 6.5.

The Redistribution
of Heat on the Globe

We stated above that the temperature of any part of the earth remains essentially constant, when averaged from year to year. In part this is due to the balanced global heat budget. However, even with such a global balance, we still have a regional imbalance, since, in terms of interactions with space, there is a net annual heat gain in lower latitudes and a net annual heat loss in higher latitudes. Were this to continue unabated, the poles would eventually cool nearly to absolute zero, and the waters of the equatorial regions would boil into space. Obviously, heat is redistributed on earth to offset the latitudinal disparity in heat absorption. This is accomplished in two major ways: through atmospheric circulation and through ocean currents. We will focus on the role of atmospheric circulation here,

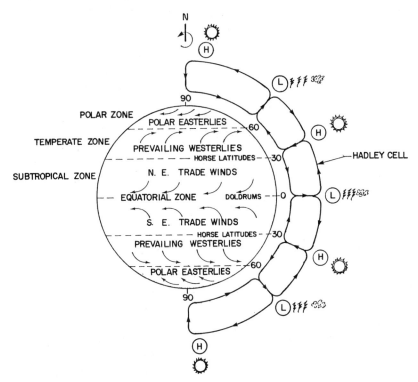

FIGURE 6.6. Idealized depiction of atmospheric circulation. See text for fuller explanation. (Adapted from Fox, W. T. 1983 *At the sea's edge.* Englewood Cliffs, N.J.: Prentice-Hall.)

saving detailed discussion of ocean currents for the next chapter. However, before we can fully appreciate the heat-redistribution function of the atmosphere, we need to understand something of the general workings of atmospheric circulation.

The Circulation of the Atmosphere

Atmospheric circulation is essentially the movement of air masses above the earth's surface. Ultimately, this circulation is driven by solar energy. Let's examine how it works.

Solar heating in low latitudes warms atmospheric air, thereby lowering its density. This light air then rises, moistened by high evaporation rates at the equator. The ascending air expands, cools again, and condenses, releasing precipitation. The cool, now dry air in the upper atmosphere spreads north and south. Its density increases as its temperature drops. Eventually, it sinks again, returning to the surface of the earth as a relatively dry air mass at approximately 30 degrees north and south latitude. Not surprisingly, most deserts are concentrated in those latitudes. This rising, spreading, and sinking again of air masses between the equator and the thirtieth parallels constitutes a heat-driven convection cell for atmospheric air, much like the hypothetical convection cells proposed for materials in the earth's mantle (described in Chapter 4). It is called a *Hadley Cell* after the meteorologist who first described it.

Once descending air reaches the surface again at the 30th parallels, it moves north and south toward the poles and equator. Because of compression it warms again slightly, which, in turn, allows it to hold more moisture. Heat and moisture absorbed from the surface of the earth reduce the density of the air mass (water molecules are lighter than other atmospheric gases). Consequently, at 60 degrees north and south, the air rises again. Thus, another convection cell is formed between the 30th and 60th parallels. In the same fashion the rudiments of still a third convection cell are evident between the 60th parallels and the poles.

The simple circulation pattern described thus far is depicted in Figure 6.6. On a nonrotating earth which lacked interfering topographic features, probably only a single convection cell, extending from equator to poles, would exist in each hemisphere. The prevailing winds would be established by the equatorward flow of air along the surface of the earth in both hemispheres. Consequently, winds would be northerly (i.e., from the north) in the northern hemisphere and southerly in the southern hemisphere. However, atmospheric circulation is complicated by the earth's rotation and the existence of continents. Let's first consider the effect of rotation.

Objects in Motion on a Rotating Earth: The Coriolis Effect

To an observer on earth, an object moving on or above the surface of the earth does not proceed in a straight line but is "deflected"—to the right in the northern hemisphere and to the left in the southern hemisphere. While this may be inherently difficult to accept, it has important practical consequences. Suppose you are an airline pilot flying a route from Mobile, Alabama to Milwaukee, Wisconsin. Both cities lie on the same line of longitude (88 degrees west), 1200 kilometers apart, so you set a course of due north at an air speed of 1200 kilometers per hour (km/hr). After an hour you see the lights of a municipal airport and initiate a landing pattern, fulling expect-

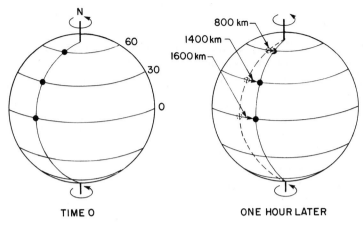

FIGURE 6.7. Because the earth is a sphere, the angular velocity of points on a rotating earth decreases with latitude in a non-linear fashion. (From Thurman, H. V. 1981. *Introductory oceanography.* 3rd ed. Columbus, Ohio: Merrill.)

ing to set down in Milwaukee a few minutes later. Immediately after establishing contact with air traffic controllers you realize something is wrong. The airport spread out below you is not Milwaukee but Detroit, Michigan, some 250 kilometers further east. How could you have made such a mistake?

Your error was in failing to take into account the earth's rotation. As our globe spins counterclockwise on its axis, every object attached to the earth moves to the east at a velocity which is dependent on latitude. An object on the equator will travel nearly 42,000 kilometers over the course of a day (i.e., the circumference of the earth at the equator). Thus, the *angular velocity* (i.e., rotational velocity) of an object at rest at the equator is approximately 1600 km/hr (Figure 6.7). However, a stationary object at 30 degrees north or south latitude has an angular velocity of only 1400 km/hr, since the circumference of the earth along that line of latitude is only 33,600 kilometers. At 60 degrees, angular velocity is down to 800 km/hr, and at the poles it is zero, since an object there undergoes no horizontal displacement on a rotating earth. There are two important points to be noted here. First, angular velocity progressively decreases with increasing latitude. Second, the rate of decrease is not uniform but, for equal latitudinal increments, increases toward the poles. Thus, a 30-degree change of latitude between the equator and 30 degrees north or south results in a 200 km/hr reduction in angular velocity, but a change from 30 to 60 degrees results

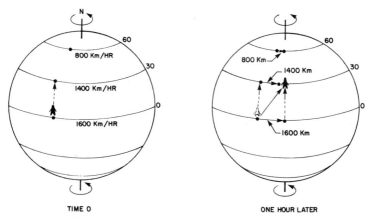

FIGURE 6.8. A plane moving to the north in the northern hemisphere undergoes an apparent deflection to the right because its eastward component of velocity (due to the earth's rotation), at the point of departure, is greater than that of its destination. It can be similarly demonstrated that any object in motion in the northern hemisphere undergoes an apparent deflection to the right.

in a reduction of 600 km/hr. Between 60 and 90 degrees the reduction is 800 km/hr.

These differences in angular velocity have important consequences for objects moving on or above the earth. To illustrate this, let's return for a moment to our airplane. Imagine that our departure point had been a city on the equator and our destination an airport due north, 30 degrees above the equator. When we left the equator, we would have been moving not only north but also east at 1600 km/hr. However, our destination site moves eastward at only 1400 km/hr. Therefore, as we crossed the 30th parallel, we would be approximately 200 kilometers east of the destination (Figure 6.8). It will appear to an earthbound observer as if we have been deflected to the right of our intended direction of motion. Of course, to an observer in space, it will be clear that we have gone in a straight line; it is the earth itself that has moved as it rotates on its axis.

If our departure point had been a city at 30 degrees north latitude and our destination an airport due north at 60 degrees, we would cross the 60th parallel at a point approximately *600* kilometers east of our destination (Figure 6.8). Therefore, we would again undergo an apparent deflection to the right of our intended motion, but its magnitude would be much greater.

This apparent deflection of a moving object on a rotating earth is called the *Coriolis effect* after the French physicist, G. G. de Coriolis, who first described it in 1824. If we had not corrected for it

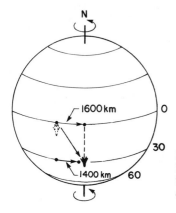

FIGURE 6.9. Objects moving in the southern hemisphere undergo an apparent deflection to the left. See Figure 6.8 and text for explanation.

(by steering to the west), our plane would have landed in Detroit instead of Milwaukee. Because latitudinal differences in the earth's angular velocity increase progressively with latitude, the magnitude of the Coriolis effect also increases with latitude.

Thus far we have restricted our examples to objects moving to the north in the northern hemisphere. An object moving to the south above the equator will also undergo an apparent deflection to its right, because its destination at a lower latitude will always have a greater angular velocity to the east than the object, which departed from a higher latitude. Thus, by the time the object crosses the latitude of the destination, the destination point will be some distance further east than the object; the object will appear to have been deflected to its right. Similarly, it can be demonstrated that objects in the southern hemisphere undergo a Coriolis deflection to the *left* (but still to the east) of their intended motion (Figure 6.9).

What about objects moving only to the east or west? Do they undergo a similar deflection? To answer this requires some understanding of the interaction between gravitational and centrifugal forces on earth. The earth is not perfectly round but is somewhat flattened into an oblate spheroid whose diameter through the equator is about 25 kilometers larger than the diameter through the poles. This distorted shape is due to a compromise between gravitational force, which tends to pull the earth into a sphere of uniform diameter, and centrifugal force, resulting from the earth's rotation, which "pushes" the earth's mass toward and outward from the equator. (Similarly, the spinning of a merry-go-round tries to propel a rider outward from the center of rotation.)

The slightly larger diameter at the equator means that the poles are 25 kilometers "downhill" from the equator. Therefore, in the absence of counteracting forces, an object on the earth's surface would tend to move toward the poles. However, centrifugal force resulting from the rotation of the earth tends to push things toward the equator. The two effects are equal for an object at rest, and the object remains stationary on the surface of the earth.

Complications arise when an object moves along a line of latitude. Since the magnitude of centrifugal force is directly related to the magnitude of angular velocity, but gravitational force is not related to the speed of a moving object, gravitational and centrifugal forces will no longer be balanced for an object moving relative to the earth's surface. If an object proceeds to the east, it moves faster than the angular velocity of the earth at that latitude. Therefore, centrifugal force exceeds gravitational force acting on the object, and it will be deflected toward the equator. In the northern hemisphere the deflection will be to the right and in the southern hemisphere to the left (Figure 6.10).

Conversely, an object moving to the west will experience a centrifugal force which is less than that for a stationary object at that latitude. Gravitational force will now be greater than centrifugal force, and the object will be deflected toward the pole. Again, the deflection will be toward the right in the northern hemisphere and toward the left in the southern hemisphere (Figure 6.10). An object moving along the equator itself will undergo no deflection, since centrifugal force is always directed toward the equator.

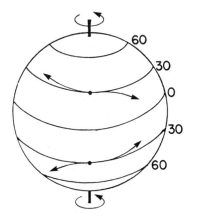

FIGURE 6.10. Objects moving to the east or west on the earth undergo an apparent deflection to the right in the northern hemisphere and to the left in the southern hemisphere. There is no deflection of objects moving along the equator. See text for full explanation.

FIGURE 6.11. Seasonal air circulation patterns are established by differential heating between land masses and the ocean. In this illustration, land masses heat up more rapidly than the water in summer. Warm, rising air creates low-pressure cells over the land, which draw in moisture-laden air from higher-pressure regions offshore. This principle is responsible for the characteristic afternoon sea breezes of coastal regions in summer. In winter, the pattern is reversed. (Adapted from Fox, W. T. 1983. *At the sea's edge.* Englewood Cliffs, N.J.: Prentice-Hall.)

The Coriolis Effect
and Atmospheric Circulation

Let's now examine how atmospheric circulation is influenced by the Coriolis effect. Recall the idealized flow of air masses at the earth's surface depicted in Figure 6.6. Between 30 degrees and the equator, surface air moves back toward the equator; between 30 and 60 degrees of latitude it moves toward the poles. However, that pattern was described for a nonrotating earth. How is it altered on a rotating globe?

Like any object moving on or above the surface of the earth, moving air is deflected by the Coriolis effect. Air masses flowing toward the equator are diverted toward the west (to the right in the northern hemisphere and to the left in the southern hemisphere), while those flowing toward the poles are moved toward the east. The amount of deflection increases toward the poles. This establishes the pattern of prevailing global surface winds depicted in Figure 6.6. Note that high-pressure cells are associated with descending air masses and low-pressure areas mark regions of rising air. Air masses might be expected to follow a pressure gradient—that is, move from the center of a high-pressure area to the center of a low-pressure region. However, the path of that movement is modified by the Coriolis effect. Since winds emanating from a high-pressure center in the northern hemisphere get deflected to the right, circulation around a high follows a clockwise pattern. In the southern hemisphere, air rotates counterclockwise around a high. Similarly, air moving into a low-pressure area gets deflected so that circulation moves counterclockwise around a low in the northern hemisphere and clockwise in the southern hemisphere.

The Effects of Land Masses on Atmospheric Circulation

In addition to the rotation of the earth, the presence of continents modifies atmospheric circulation. One of the most important effects of large land masses is to establish seasonal patterns of high- and low-pressure centers. Because of water's large thermal inertia, it changes temperature far more slowly than land. Generally, the land will be colder than the ocean in the winter and warmer in the summer. Warm air rising off continents in the summer creates low-pressure cells which draw in air masses from relatively high-pressure areas offshore. This accounts for the predominance of "sea breezes" (blowing from sea toward land) in coastal areas in the summer (Figure 6.11). Conversely, cold, dense air sinks over land in the winter, creating characteristic high-pressure centers. Consequently, winter air masses tend to move from land toward the oceans. These patterns are particularly evident in Asia, where the seasonal heating and cooling of the large continental land mass causes the seasonal *monsoons* (see also Chapter 7). When moist oceanic air masses move in toward the continent during the summer, the monsoons are wet and stormy. Continental air moving into the Indian Ocean during the winter establishes the dry monsoons of that season (Figure 6.12).

The Effects of Atmospheric Circulation on Global Heat Distribution

With this background on atmospheric circulation we can now address its role in global heat redistribution. We know that, over the long run, the poles aren't cooling down and the equator isn't getting warmer, despite the significant disparity in absorption of solar energy by the two regions. We have only to refer back to the atmospheric convection cells depicted in Figure 6.6 to begin to understand why. The equatorial region acts as a giant heat pump, spewing massive quantities of warm, moist air into upper altitudes.

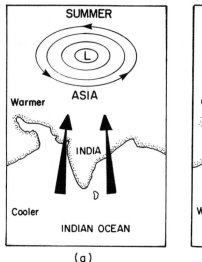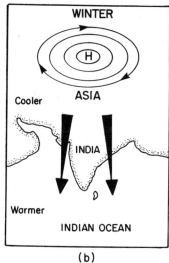

FIGURE 6.12. The seasonal monsoons of Asia. (a) The "wet" monsoons of summer. (b) The "dry" monsoons of winter. For explanation, see text.

Some of this heat is sensible heat and results in direct warming of the atmosphere. Most of it, however, is latent heat, representing the heat stored in evaporated sea water from tropical regions. A careful observer in the tropics could see this process in action. Thousands of small clouds form large, scattered clusters in tropical areas. Eventual condensation of the moisture absorbed by a cloud cluster releases precipitation to the tropics and huge amounts of heat to the upper atmosphere, some 12 kilometers above the earth's surface. The air, now cooler and drier, drifts slowly poleward at high altitudes, eventually sinking to earth again in the subtropical regions. This transport of air from the equator toward higher latitudes helps to redistribute heat on the globe.

Now let us consider more specific influences of the oceans on weather and climate.

THE INFLUENCE OF THE OCEANS ON WEATHER AND CLIMATE

There is a tendency by some people to use the terms *weather* and *climate* interchangeably, but they represent quite different time scales. Weather may be considered a temporary and localized state of the atmosphere, measured on a time scale of hours to days. Climate refers to a long-term (involving periods of months to years or more) average of weather conditions on a larger geographic scale.

The oceans have a significant influence on both weather and climate which is manifested in a number of ways. We will address some of these effects herein.

Coastal Climates

The oceans significantly modify coastal climates because of the high heat capacity and thermal equilibrium of water (see Chapter 5). Consequently, coastal regions are said to have *maritime climates*, in contrast to *continental climates*, which are far from the seas' influence. Maritime climates characteristically cool down and warm up more slowly and have a smaller annual range of temperature than continental climates.

The Influence of Ocean Circulation

The circulation patterns of the seas have a significant effect on global climate in that ocean currents transport large quantities of heat from equatorial regions to polar latitudes. For example, every second the Gulf Stream moves up to 90 million cubic meters of relatively warm water from the Caribbean toward the North Atlantic. Large eddies which spin off the Gulf Stream and other major warm currents on the earth are also believed to contribute substantially to redistribution of heat from low to high latitudes. The dynamics and effects of the ocean currents will be discussed in greater detail in Chapter 7.

Large-Scale Anomalies of Ocean Temperature and Overlying Air Pressure

The extraordinary meteorological events in the winter of 1982-1983, described in the beginning of this chapter, appear to have been attributable to a pattern of fluctuations in precipitation, air pressure, and temperature in the Pacific called the *Southern Oscillation*. Related to this pattern is the periodic occurrence of an abnormally warm "current" in the equatorial regions of the eastern Pacific called *El Niño* ("the infant" in Spanish, so named because the event often occurs around Christmas). The combined events are referred to as the *El Niño/Southern Oscillation* (ENSO). Years of research on the ENSO phenomenon, capped by a flurry of activity after the onset of the 1982-1983 event, have resulted in the tentative establishment of a hypothesis which attributes some global climate fluctuations to the interacting effects of sea-surface temperature (SST) at the equator, precipitation in the tropics, and patterns of circulation in the *jet stream* (the narrow, swift, and meandering circumglobal winds which girdle the earth at a height of about 10 kilometers).

Under usual circumstances, surface-water temperatures in the western equatorial Pacific are several degrees warmer than those in the east. This strong SST gradient is associated with a characteristic atmospheric circulation pattern at the equator which includes the equatorial extension of the trade winds, which blow toward the west. The equatorial western Pacific experiences relatively low atmospheric pressure and high precipitation, while Pacific equatorial regions east of the International Date Line (the 180° meridian, in the central Pacific) have higher atmospheric pressures and are comparatively dry. Sea level in the tropical western Pacific is also higher (by up to a half meter) than that in the east, apparently because the driving force of the easterly trade winds piles up the waters there.

When a "normal" ENSO episode occurs, the east-west equatorial SST and atmospheric pressure gradients are sharply reduced. The western equatorial Pacific experiences droughts, and eastern areas become wetter as the region of heavy precipitation shifts eastward. Normally dry islands in the central Pacific experience torrential downpours. The easterly equatorial trade winds (which are characteristically quite strong prior to normal ENSO events) slacken and may be replaced by westerly winds (blowing from the west to the east). Accompanying these changes is a massive surge of warm equatorial surface waters from the west to the east until, at their eastward extension off the west coast of South America, the warm waters overlie the normally cool surface waters. This event effectively prevents the upwelling of cold, nutrient-rich deep waters which usually occurs off the west coast of Peru (see Chapter 7). Without this upwelling the surface waters become depleted of nutrients, causing temporary declines in the productivity of these fisheries-rich waters.

The 1982-1983 ENSO episode was unusual in that it began much earlier in the year than normal, was much stronger, persisted much longer, and was not preceded by abnormally strong trade winds. Further, its effects on global climate appear to have been of a

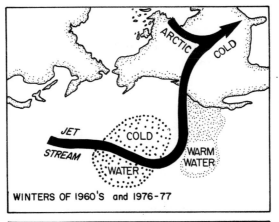

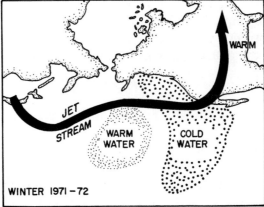

FIGURE 6.13. The effect of anomalously warm and cold patches of water in the Pacific Ocean on the path of the jet stream. (From National Science Foundation. 1982. *Report of the decade: The International Decade of Ocean Exploration,* p. 22.)

severity that is unprecedented, at least in this century. The causes of the "abnormalities" of this most recent episode and, indeed, of the ENSO phenomenon in general, are not yet well defined, underscoring the need for further research on the important interactions between the oceans and the atmosphere, not only in the tropics but in other areas of the world as well. To this end, a major U.S. research program on the ENSO phenomenon is now being planned.

Anomalously warm or cold "pools" of surface waters in the oceans of the world, including those associated with ENSO episodes, are also implicated in shifts in the path of the jet stream (Figure 6.13). This phenomenon is believed responsible for major global climatological events, including unusually warm or cold winters in different regions of North America.

Hurricanes

Despite their dangerous reputation, hurricanes (and their Pacific equivalents, typhoons) serve an extremely useful purpose by releasing excess heat energy that has accumulated during the summer in low latitudes. The amount of energy involved is astonishing—a single hurricane may pack the power of 400 twenty-megaton hydrogen bombs and contain more energy than the entire amount industrially generated in the United States in a 20-year period. Though their unpredictable path may bring them ashore, with potentially destructive results, they appear to play a major role in maintaining the global heat balance described above.

The Oceans and Carbon Dioxide

There has been considerable recent publicity about a potentially catastrophic warming of the earth due to the accelerating accumulation of carbon dioxide in the atmosphere, largely attributable to fossil-fuel burning. A certain amount of atmospheric CO_2 is crucial to life on earth for photosynthesis and because of the previously described "greenhouse effect" (the absorption and earthward reradiation, by atmospheric carbon dioxide and water vapor, of infrared radiation emitted by the earth). Without this effect, the earth might be plunged into subfreezing temperatures, and the oceans might be frozen solid. However, it is now widely believed that if the current trend of rising global atmospheric CO_2 concentrations continues, significant global warming could occur by sometime in the next century. According to estimates of Roger Revelle, former Director of the Scripps Institution of Oceanography, a doubling in atmospheric CO_2 is likely to occur in the next 100 years. This would probably raise average global atmospheric temperatures by 2 to 3 ° C, with larger increases expected in polar regions. The ultimate effects of such warming cannot be predicted with certainty, but Revelle anticipates significant melting of glacial and sea ice, with a resultant rise in sea level of at least 60 centimeters in the next century, about four times the current rate of rise.

Since the oceans already contain 53 times as much carbon dioxide as the atmosphere, one might expect that they would easily absorb the comparatively small quantities introduced anthropogenically (i.e., by humans). However, the oceans' capacity for CO_2 absorption is limited. According to best estimates, the seas have probably absorbed between 30 to 40 percent of the carbon dioxide introduced into the atmosphere by human activities. Terrestrial forests and other components of the biosphere have probably taken up another 10 to 30 percent. Thus, the remainder has accumulated in the atmosphere.

Oceanographic Studies of Climate

The decade of the 1970s saw an increasingly important emphasis on oceanographic investigations of climate, an emphasis that continues to the present. Several studies supported by the IDOE (see Chapter 1) deserve particular mention here.

The North Pacific Experiment (NORPAX) was designed to examine air-sea interactions in the Pacific, focusing particularly on the existence, proposed by Jerome Namias, of large (up to 2000 kilometers in diameter) patches of anomalously warm or cold water in the Pacific. Namias had hypothesized that such patches affected weather and climate in North America (see earlier sections of this chapter). By monitoring sea-water properties and sea-surface temperatures through ship-based instrumentation and satellite-tracked buoys, the researchers confirmed the existence of the thermal patches. The patches' effects on global climate are still being investigated. NORPAX investigators also conducted research on the ENSO phenomenon, described above.

The hypothesized effects of large ocean eddies (see Chapter 7) on global climate helped to stimulate a series of studies of these phenomena in the 1970s. The research projects, called MODE-1, MODE-1 Extension, and POLYMODE, helped to define the role of the Gulf Stream in producing eddies in the western North Atlantic and began the complex process of elucidating the influence of these eddies on climate and weather.

The International Southern Ocean Studies (ISOS) program was also conducted during the 1970s. It eventually involved scientists from 12 countries, whose purpose was to gain a comprehensive understanding of the Antarctic Circumpolar Current (see Chapter 7), in part because of the hypothesized effect of its role in global climate dynamics. An important result of these studies was the discovery of large Southern Ocean eddies which transport significant quantities of heat from lower latitudes toward the South Pole.

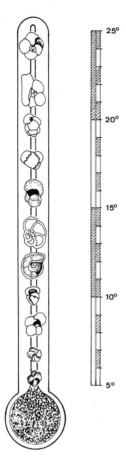

FIGURE 6.14. A "paleothermometer." Fossil remains of marine microorganisms indicate the approximate water temperature at the time they were alive, because different species had different temperature tolerances. (From National Science Foundation. 1982. *Report of the decade: The International Decade of Ocean Exploration*, p. 28.)

The study of ancient climates, through paleontological investigations of sea-floor sediments, was the focus of a major research program of the 1970s. The project, called Climate: Long-Range Investigation, Mapping, and Prediction (CLIMAP), attempted to reconstruct the climate history of the past million years, in part by analyzing the skeletons of long-dead microscopic marine organisms, including foraminifera and radiolarians (see Chapters 3 and 9), and diatoms and coccolithophores (see Chapter 9). By taking cores of the sedimentary profile of the sea floor (see Chapters 2 and 4) and using established dating techniques, investigators were able to draw correlations between the age of the sediments at a particular depth in the core and the taxonomic identity of the associated organisms. By knowing the environmental requirements for related species of modern organisms and assuming these requirements were the same for their ancestors, scientists can infer the climate conditions at the time the fossils were alive (Figure 6.14).

Today, oceanographic investigations of climate are relying more and more on remote-sensing techniques (see Chapter 1). The potential climate-related applications of satellite sensing devices include fine-scale mapping of sea ice cover (using synthetic-aperture radar), which may provide early warning of the enhanced "greenhouse effect"; analysis of temperature structure in the oceans' surface waters (an accuracy to within 1.3 ° C and a resolution of 150 kilometers are possible); measurements of ocean winds (particularly helpful in studying the ENSO phenomenon); tracking of ocean eddies (in part through infrared imagery, which highlights temperature contrasts between eddies and surrounding waters) and currents (aided by color-scanning detection of concentrations of chlorophyll, which are an indicator of phytoplankton populations within a current); and extremely precise gravitational measurements of sea-surface height (a technique called *spaceborne altimetry*).

SUGGESTED READINGS

BRETHERTON, F. P. 1981. Climate, the oceans, and remote sensing. *Oceanus* 24:48-55.

CALDER, N. 1974. *The weather machine.* New York: Viking Press.

CANE, M. A. 1983. Oceanographic events during El Niño. *Science* 222:1189-94.

HALPERN, D., HAYES, S. P., LEETMAA, A., HANSEN, D. V., AND PHILANDER, S. G. H. 1983. Oceanographic observations of the 1982 warming of the tropical eastern Pacific. *Science* 221:1173-75.

LAMB, H. H. 1968. *The changing climate: selected papers.* London: Methuen and Co.

MACLEISH, W. H., ed. 1978. The oceans and climate. *Oceanus* 21(4). (A collection of articles.)

NAMIAS, J. 1981. The weather: from sea to sky. *Oceans* 14:44-53.

RASMUSSEN, E. M., AND WALLACE, J. M. 1983. Meteorological aspects of the El Niño/Southern Oscillation. *Science* 222:1195-1202.

REVELLE, R. 1982. Carbon dioxide and world climate. *Scientific American* 247:35-43.

REVELLE, R. 1983. The oceans and the carbon dioxide problem. *Oceanus* 26:3-9.

SOBEY, E. 1980. The ocean-climate connection. *Sea Frontiers* 26:25-30.

Chapter Seven

Ocean Circulation:
Currents

THE STEADY MARCH OF OFFSHORE SWELLS, THE CRASH OF BREAK-
ers on a beach, and the rhythmic advance and retreat of global tides
involve wave motion on the sea surface. Relatively little water displacement occurs. In contrast, the great ocean currents transport
massive quantities of water over vast distances. This transport may
be horizontal or vertical, depending on the nature of the driving
force which sets the currents in motion.

There are two principal causes of ocean currents: the atmospheric winds and density differences between water masses. As
winds sweep across the ocean, they exert a frictional force on the sea
surface, horizontally displacing the affected water mass. This establishes the *wind-driven circulation.* The speed and direction of
wind-driven currents are modified by a number of factors, including
the Coriolis effect, the shape of ocean basins, and the presence of
land masses.

Density differences between water masses are responsible for
the *thermohaline circulation* (see page 150). This circulation has a
significant vertical component and ultimately derives from the sinking of surface waters in response to increased density. Thermohaline circulation is responsible for a slow overturning of the oceans'
waters. Without such mixing, which recharges the surface waters
with nutrients, the seas' productivity would be vastly diminished.

The objective of this chapter is to describe some of the important aspects of ocean currents, including the major factors which establish the currents and influence their behavior.

THE WIND-DRIVEN CIRCULATION

Wind-driven currents are set in motion by moving air masses and
shaped by the earth's rotation and the configuration of the continents. Moving air stresses the sea surface and the waters move in response, but the relationship between wind and water is not
straightforward. Because of the Coriolis effect (see Chapter 6),
water masses in the seas' upper layers flow at an angle to the wind
direction. Let's examine this response.

Imagine a calm, stationary sea surface in the northern hemisphere. A gentle breeze begins to blow, gradually intensifying over
time. As the wind drives against the sea surface, a column of water,
whose depth is a function of wind speed, duration, and fetch (see
Chapter 8), moves in response. The column does not move uni-

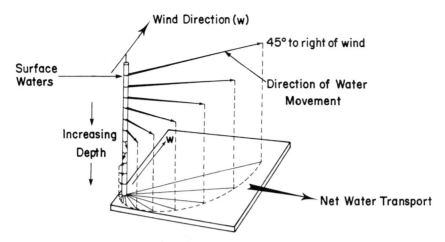

FIGURE 7.1. Depiction of the Ekman spiral in the northern hemisphere. Successively deeper layers of water move successively further to the right of the wind direction at successively slower velocities. The surface current is 45 degrees to the right of the wind and the net water transport is 90 degrees to the right of the wind. (From Sverdrup, H. U., Johnson, M. W., and Fleming, R. H. 1942. *The oceans.* Englewood Cliffs, N.J.: Prentice-Hall.)

formly. As the wind exerts a frictional stress on the surface layer, the water begins to move in the direction of the wind but is immediately acted upon by the Coriolis effect. This deflects the water motion to the right of the wind direction (to the left in the southern hemisphere). Simultaneously, frictional stress between the surface layer and the water layer below it retards the movement of the surface layer. The resultant flow of the surface layer represents a steady-state condition in which the surface water moves in a direction 45 degrees to the right of the wind direction (to the left in the southern hemisphere).

To understand the net movement of the entire column of water affected by the wind, it is helpful to picture the column as a series of water layers. By applying the same principles as those presented for the movement of the surface layer, we can show that each progressively deeper layer of water moves to the right of the layer above it and at a slightly lower velocity. If we then depict the movement of water layers as a series of directional arrows whose length corresponds to relative velocity (Figure 7.1), we see that the representation resembles a spiral staircase with successively shorter steps arranged at progressively greater angles from the wind direction. For this reason, the circulation pattern is called the *Ekman spiral*, after the Swedish physicist V. W. Ekman, who first described it

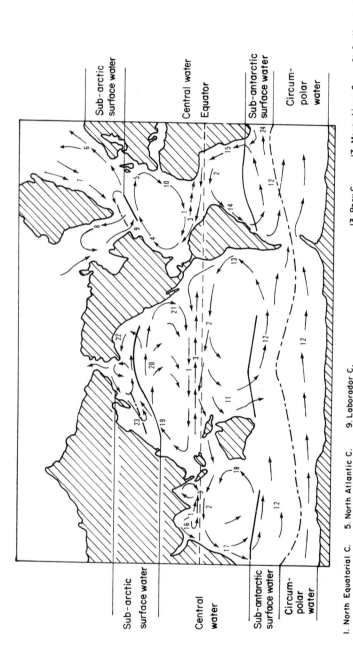

Sub-arctic
surface water

Central water

Equator

Sub-antarctic
surface water

Circum-
polar
water

Sub-arctic
surface water

Central
water

Sub-antarctic
surface water

Circum-
polar
water

1. North Equatorial C.	5. North Atlantic C.	9. Laborador C.	13. Peru C.	17. Mozambique C.	21. California C.
2. South Equatorial C.	6. Norwegian C.	10. Canary C.	14. Brazil C.	18. West Australian C.	22. Aleutian C.
3. Equatorial Counter C.	7. East Greenland C.	11. East Australian C.	15. Benguela C.	19. Kuroshio C.	23. Oyashio C.
4. Gulf Stream C.	8. West Greenland C.	12. Antarctic Circumpolar C. (West Wind Drift)	16. Somali C.	20. North Pacific C.	24. Agulhas C.

FIGURE 7.2. The major surface currents and surface water masses of the oceans. Arrows indicate the directions of water movement, heavy dark lines show subtropical convergences and heavy dashed lines depict arctic and antarctic convergences. (From Anikouchine, W. A., and Sternberg, R. W. 1973. *The world ocean.* Englewood Cliffs, N.J.: Prentice-Hall.)

mathematically (see Chapter 1). Note that, at some point deep in the water column (the 180-degree reversal depth), a layer of water will actually move in a direction opposite to that of the surface waters and 225 degrees from the wind direction.

Under ideal conditions, the Ekman spiral results in a *net transport* (*Ekman transport*) of wind-driven water in the affected water column (the *Ekman layer*) which is 90 degrees from the wind direction (to the right in the northern hemisphere and to the left in the southern hemisphere). In shallow water it will be somewhat less than 90 degrees because of the restricting influence of the ocean bottom.

The Oceanic Gyres and Geostrophic Currents

Now that we know something about the response of the oceans' surface waters to wind stress, we can consider large-scale wind-driven circulation. Refer to the global surface-water circulation patterns depicted in Figure 7.2. A characteristic feature of the major surface currents is that they describe large circular orbits, or *gyres*, within individual ocean basins. The circulation of these gyres is "anticyclonic" (i.e., clockwise in the northern hemisphere and counterclockwise in the southern). Typical of the ocean basin gyres is a major clockwise current flow in the North Atlantic called the *North Atlantic gyre.*

There may be a temptation to consider an oceanic gyre as a kind of giant whirlpool with water piled up around the outer edges and a depression in the center. In fact, topographically it is just the opposite, rather resembling a hill with its peak within the gyre (Figure 7.3). The "hills" in oceanic gyres are not immediately obvious—their peaks rise only about two meters above the base of their slopes. Nonetheless, the hills have a significant effect on ocean circulation.

What causes the mounding up of water in oceanic gyres? Ultimately, Ekman transport is responsible. Remember that the net transport of an entire column of wind-driven water is approximately 90 degrees from the wind direction and about 45 degrees from the direction of the uppermost surface currents. Consequently, there is a continual transport of surface waters toward the interior of a gyre. This *convergence* of waters results in a slight hilling up of the sea surface. The mounding up of relatively warm, low-density surface

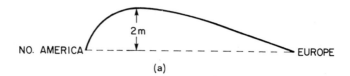

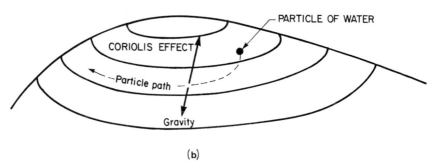

FIGURE 7.3. Topography and circulation of oceanic gyres. (a) The topographic highs of gyres tend to be offset to the west, as exemplified by the North Atlantic gyre. (b) Geostrophic flow of surface currents in oceanic gyres. Water runs around the hill because of an equilibrium established between gravity, which pulls water down the hill (a "downslope" component), and the Coriolis effect, which deflects water into the hill (an "upslope" component). (Adapted from Thurman, H. V. *Introductory oceanography*. 3rd ed. Columbus, Ohio: Merrill.)

waters in the interior of the gyres depresses the colder, higher-density waters below.

Now consider the behavior of individual water droplets on the slope of the hill. In response to gravity the droplets begin to roll back down the hill; however, once they are in motion the Coriolis effect deflects them to the right of their downhill path. The balance established between the *pressure-gradient force* (which results from the difference in pressure between two points on the hill because of the sloping sea surface) and the Coriolis effect causes the surface waters in the gyre to follow a curved path conforming to the contours of the hill (Figure 7.3) and limits the slope and height of the hill. In effect, the water runs *around* the hill.

When the Coriolis effect and the pressure-gradient force are in exact balance, they are said to be in *geostrophic balance*. Consequently, the movements of surface waters around oceanic gyres are called *geostrophic currents*. They are analogous to the atmospheric flow around high- and low-pressure systems (see Chapter 6). The geostrophic currents are responsible for the circular movement of waters in oceanic gyres and conceivably would serve to maintain the gyres for some time even if the global winds suddenly ceased.

Westward Intensification of Ocean Currents. The individual currents which comprise a specific oceanic gyre do not flow at the same speeds. Currents along the western margins of ocean basins (the *western boundary currents*) are considerably swifter and narrower than their counterparts on the eastern side. For example, the velocity of the Gulf Stream approaches 9 kilometers per hour, while the Canary Current flows south along the European coast at the comparative snail's pace of less than 1 kilometer per hour. This *westward intensification* of gyre currents results from a balance between several factors which affect the surface currents. These include the increasing magnitude of the Coriolis effect with latitude (see Chapter 6), latitudinal variations of wind strength and direction, and friction between the water currents and contiguous land masses. These factors cause the topographic highs of the oceanic gyres to be offset to the west (Figure 7.3). The western boundary currents move faster than those on the eastern side because a constant volume of water is squeezed into a narrower band along the western margins of the ocean basins (Figure 7.4).

Ekman Transport and Vertical Water Movement

We associate winds with horizontal circulation, but they often cause substantial vertical displacement of water masses. Under some cir-

FIGURE 7.4. Surface water circulation in an idealized ocean basin. Currents along the western margin are characteristically swifter and narrower than those further to the east. (From Stowe, K. 1983. *Ocean science.* 2nd ed. N.Y.: Wiley.)

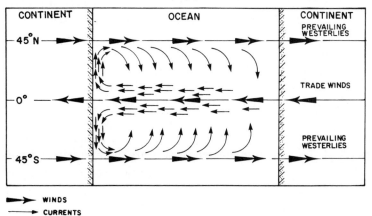

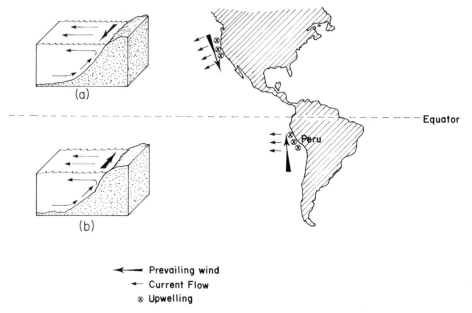

——◄— Prevailing wind
◄— Current Flow
⊗ Upwelling

FIGURE 7.5. Coastal upwelling north and south of the equator along the west coast of the Americas. (Adapted from Meadows, P. S., and Campbell, J. I. 1978. *An introduction to marine science.* Glasgow: Blackie and Son, Ltd.)

cumstances wind patterns engender a continuous rising of water from the depths to the surface, a phenomenon known as *upwelling*. A reverse vertical flow called *downwelling* occurs under other circumstances. While wind-induced vertical circulation is not widespread, it has disproportionate effects on oceanic productivity and is worth considering in some detail.

Upwelling is prevalent adjacent to continental margins and along the equator. It occurs when prevailing winds cause Ekman transport of surface waters away from the affected region. Deep waters then well up to replace them. Let's look at *coastal upwelling* first.

Coastal upwelling occurs when prevailing winds blow parallel to the coastline in a direction favoring Ekman transport of surface waters away from the coast. In the northern hemisphere this will occur when the coast is to the left of the wind direction and in the southern hemisphere when the shoreline is to the right of the prevailing wind (Figure 7.5). Deep, cold waters which rise up to take the place of the vacating surface waters are characteristically rich in the nutrients which stimulate growth of microscopic floating plants. These phytoplankton in turn fuel a highly productive ecosystem

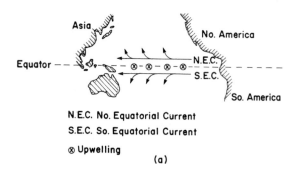

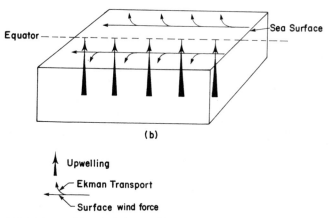

N.E.C. No. Equatorial Current

S.E.C. So. Equatorial Current

⊗ Upwelling

(a)

↑ Upwelling

↖ Ekman Transport

← Surface wind force

FIGURE 7.6. Equatorial upwelling.

which often culminates in substantial stocks of commercially valuable fish. The Peruvian herring fishery is a frequently cited example. Along the coast of Peru prevailing southeast winds transport surface waters offshore (Figure 7.5). The resultant upwelling-induced productivity has contributed to some astonishingly high fish harvests in that region. At one time nearly one quarter of the entire world fish catch came from Peruvian waters. Unfortunately, the prevailing winds fail infrequently and unpredictably (see Chapter 6). When this occurs, upwelling ceases and productivity plunges precipitously, contributing to the collapse of the fishing industry.

Upwelling also occurs along the west coast of North America because the prevailing winds are from the north along shore (Figure 7.5). In addition to enhancing productivity, cold, upwelled waters keep adjacent coastal climates comparatively cool on a year-round basis.

Equatorial upwelling occurs along the equator. The westward-blowing trade winds cause Ekman transport of surface waters away from the equator on both sides (Figure 7.6). Deep, cold water then rises up to take the place of the diverging water masses. (*Divergence* is the term used to describe the separation of water masses in the ocean, while *convergence* indicates their coming together.)

Downwelling occurs at oceanic convergences and when water is piled up along a coast because of Ekman transport toward shore (Figure 7.7).

Wind-Driven Circulation in the North Atlantic

Let us now examine wind-driven, large-scale circulation patterns in greater detail, using the North Atlantic as an example. To a large extent the circulation patterns of the major oceans are surprisingly similar, in that they characteristically involve a large central gyre and one or more secondary gyres.

The general circulation pattern of the North Atlantic is depicted in Figure 7.8. As the trade winds blow steadily from the northeast between 30 degrees north latitude and the equator, the surface waters are deflected to the right and pushed toward the interior of the mid-Atlantic gyre, where a high-pressure zone is maintained. This establishes a major, westward-flowing geostrophic current, the North Equatorial Current. Eventually, the current reaches the western margin of the North Atlantic Basin, where land masses deflect it toward the north. There it merges with a portion of the South Equatorial Current which has been deflected to the north by South America. The combined current moves along the north-

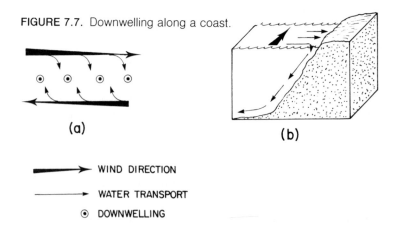

FIGURE 7.7. Downwelling along a coast.

(a)

(b)

━━━▶ WIND DIRECTION

──────▶ WATER TRANSPORT

⊙ DOWNWELLING

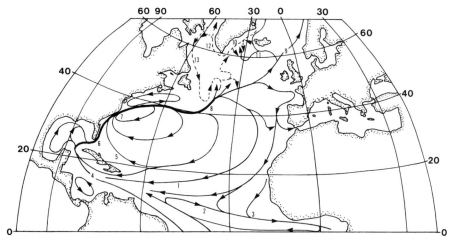

1 NO. EQUATORIAL CURRENT	8 NO. ATLANTIC CURRENT
2 EQUATORIAL COUNTERCURRENT	9 NORWEGIAN CURRENT
3 GUINEA CURRENT	10 E.GREENLAND CURRENT
4 CANARY CURRENT	11 IRMINGER CURRENT
5 ANTILLES CURRENT	12 W. GREENLAND CURRENT
6 FLORIDA CURRENT	13 LABRADOR CURRENT
7 GULF STREAM	

FIGURE 7.8. The generalized surface water circulation of the North Atlantic. Solid lines indicate warm currents, and dashed lines indicate cold currents. (Adapted with permission from Tchernia, P. 1980. *Descriptive regional oceanography*. Oxford: Pergamon Press.)

west coast of South America. Most of this current flows below the Caribbean Islands as the Caribbean Current. This loops into the Gulf of Mexico, then back out around the south coast of Florida, where it is known as the Florida Current. The Florida Current then surges along the east coast of North America, ultimately becoming the Gulf Stream.

The Gulf Stream is eventually deflected to the right by the topography of the sea floor. South of the Grand Banks of Newfoundland, the Gulf Stream becomes a midlatitude current which flows eastward toward Europe as the North Atlantic Current. When the North Atlantic Current reaches the eastern margin of the North Atlantic Basin, part of it flows north and then west. Another branch moves into the Norwegian Sea as the Norwegian Current. The rest flows south along the coast of Europe as the Canary Current. Eventually, the Canary Current rejoins the North Equatorial Current, completing the closed loop of the North Atlantic gyre.

Between the North and South Equatorial Currents flows a narrow, eastward-flowing current called the *Equatorial Counter-current*. A well-established Equatorial Countercurrent also occurs in the Pacific Ocean.

In addition to the equatorial countercurrents, other eastward-flowing currents move along the equator in the Pacific, Atlantic, and Indian (seasonally) Oceans. However, these are subsurface currents, generically referred to as *equatorial undercurrents*. The Pacific equatorial undercurrent, known as the Cromwell Current, is the best known.

Only about 250 meters thick and 250 kilometers wide, equatorial undercurrents reach speeds of up to 5 kilometers per hour. Their causes are not yet completely understood, though it has been hypothesized that they represent an eastward return flow of water which is piled up along the western margins of ocean basins by the North and South Equatorial Currents. If so, their flow is presumably attributable to the pressure gradient caused by the slope of the sea surface and is not modified by the Coriolis effect, which is non-existent at the equator (see Chapter 6).

Of all of the surface currents in the North Atlantic, the Gulf Stream is easily the most familiar and, perhaps, the most important. Let's examine it in closer detail.

THE GULF STREAM. "Stream" is too humble a term for the mighty river in the sea which surges past the heavily populated United States Atlantic seaboard. The Gulf Stream is the most glamorous and best-studied component of the North Atlantic circulation. It is fully deserving of this attention. Born in the tropics, it moves up to 90 million cubic meters of water every second, 25 times the combined flow of all the rivers in the world. Its boundaries are not constrained by banks or levees, but its presence is immediately obvious to the yachtsman who crosses the sharp transition between the languid, greenish, and relatively turbid continental slope waters and the narrow, swift, clear, deep-blue, and contrastingly warm current which constitutes the Gulf Stream. Only 50 to 75 kilometers wide, and moving at 3 to 10 kilometers per hour, it effectively divides the relatively cold, nutrient-enriched waters of the Atlantic coast from the warm, nutrient-poor waters in the interior of the North Atlantic gyre. Far deeper than any terrestrial river, the Gulf Stream's lower limits may sometimes be detected to the depth of the sea floor. In contrast to the western edge, the eastern boundary of the Stream is diffuse; the current gradually merges into the Sargasso Sea within the North Atlantic gyre.

An important aspect of Gulf Stream circulation is its spatially varying volume transport. As the ocean current knifes past southern Florida, it moves about 30 million cubic meters of water every second. By the time the Stream passes the latitude of Bermuda, its transport has increased to 150 million cubic meters per second. However, by the time the Gulf Stream meanders past the Grand Banks of Nova Scotia, its volume has been significantly reduced again. Part of this change is due to the recirculation of water within the Sargasso Sea. However, another contributing factor is an important feature of the Gulf Stream system—*Gulf Stream rings.*

GULF STREAM RINGS. Swift as the Gulf Stream may be, its path is not sure. Like an immense, indecisive snake it meanders sinuously and unpredictably as it flows toward higher latitudes. In its vacillating progress it frequently loops back on itself, in the process pinching off large, ring-shaped progeny, called Gulf Stream rings or *eddies* (Figure 7.9).

FIGURE 7.9. Infrared photograph of the western North Atlantic. The broad dark band running from southwest to northeast is the Gulf Stream. Note a warm-core ring in the right center of the photograph and another ring being pinched off in the center. (Photo courtesy of the Woods Hole Oceanographic Institution.)

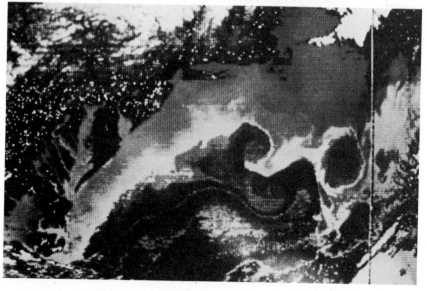

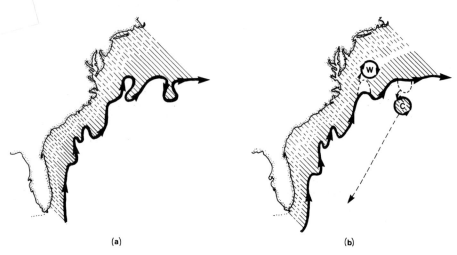

FIGURE 7.10. The formation of Gulf Stream rings. W = warm-core rings.
C = cold-core rings. Shaded areas = slope waters. Clear areas = Sargasso
Sea waters.

Gulf Stream rings and their counterparts in several other
ocean-current systems have been known of for over 50 years, but
only recently has it been discovered that they may have important
influences on the climate of the northern hemisphere and the ecology of the northern seas. This discovery has spawned a major research effort over the past decade or so. While we still have much to
learn about these important components of ocean circulation, we
have gained considerable understanding of their origin, behavior,
and effects.

We have learned that the rings have lives of their own, moving
independently of the Gulf Stream with a rotational velocity of about
4 kilometers per hour. Most undergo a variable lateral transport,
traveling as much as 4 kilometers a day. Some seem not to move
much at all for weeks at a time.

We have also learned that there are two kinds of Gulf Stream
rings, created when the meandering path of the current loops back
on itself. Loops billowing out to the south will spawn eddies which
move toward the southwest, inside the North Atlantic gyre (Figure
7.10). These eddies spin counterclockwise and contain a central
core of relatively cold water from north of the Gulf Stream, trapped
when the ring was created. Consequently, they are called *cold-core
rings*. Cold-core rings are characteristically 150 to 200 kilometers in
diameter and may be much larger. They often descend all the way
to the sea floor.

Because their cold central temperatures mean that their density is greater than that of surrounding Sargasso Sea water, the interior of the rings is depressed as much as a half meter below the waters around them. As they move southwest toward the northern Caribbean, they gradually lose energy and individuality. Some die a slow death, bit by bit merging with surrounding waters. Many, however, appear to live long enough to be reabsorbed by the Gulf Stream when their journey brings them back to the current's outer limits in the northern Bahamas (Figure 7.10).

Warm-core rings are formed when a northward convolution of the Gulf Stream closes back on itself, trapping a parcel of relatively warm Sargasso Sea water in the process (Figure 7.10). These clockwise-rotating rings are confined to the "slope waters" of the North American continent, west of the Gulf Stream. Slightly smaller than their cold-core counterparts, they average 100 to 200 kilometers in diameter and descend up to two kilometers below the sea surface. Like cold-core rings, their path of travel is to the southwest. Eventually, most of them are also reabsorbed back into the Gulf Stream. They have a mean lifetime of four and a half months, but many last less than three months, and some live as long as a year. Because the center of warm-core rings is less dense than the surrounding cold water, their surfaces are elevated as much as a half meter above the level of slope waters.

There is considerable interest in the effects of Gulf Stream rings on the climate, ecology, water quality, and fisheries of the North Atlantic. Eddies apparently result in a significant transport of cold-slope waters into warmer Sargasso Sea waters and vice versa. While exact volumes involved are still uncertain, it is likely that this transport contributes significantly to the redistribution of global heat.

Just as rings enclose and transport thermal masses, so do they carry the chemical and biological properties of the waters which they entrap. Thus, significant amounts of nutrients from nutrient-enriched slope waters may be moved into the nutrient-poor waters of the Sargasso Sea, perhaps locally stimulating productivity in the process. Similarly, plankton, larvae, and fish may be transported from their home waters to regions where they would not ordinarily survive, temporarily buffered from the alien waters by the whirling remnant of the Gulf Stream. It has even been suggested that if the mini-ecosystem of a ring changes slowly enough and persists long

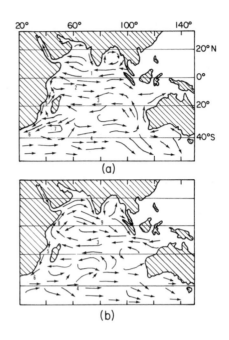

I	NO. EQUATORIAL CURRENT	6	AGULHAS CURRENT
2	EQUATORIAL COUNTER CURRENT	7	RETURNING CURRENT
3	SO. EQUATORIAL CURRENT	8	SOMALI CURRENT
4	WEST WIND DRIFT	9	DRIFT CURRENT of the
5	MOZAMBIQUE CURRENT		SOUTHWEST MONSOON

FIGURE 7.11. Seasonal surface-water circulation in the Indian Ocean. (a) Circulation in February. (b) Circulation in August. (Reprinted with permission from Tchernia, P. 1980. *Descriptive regional oceanography.* Oxford: Pergamon Press.)

enough to allow several generations of enclosed species to develop, the species may adapt, through natural selection, to the environmental conditions of the new region.

Our interest in Gulf Stream rings extends to their possible adverse effects. Their ability to trap and transport heat, chemicals, and nutrients has caused concern that they may similarly move and disperse dangerous pollutants from localized ocean dump sites if their unpredictable course carries them into such regions. An interesting but fortuitous related incident occurred after the grounding of the *Argo Merchant* off Nantucket Island in 1976. Much of the resultant oil spill was drawn into a warm-core ring which subsequently distributed it over a wide region offshore, preventing it from dispersing on shore as had been originally feared.

Indian Ocean Circulation

The surface circulation in the Indian Ocean is noteworthy because it is influenced strongly by the contrasting thermal characteristics of the large Asian land mass and the waters offshore (see Chapter 6). This contrast establishes distinct seasonal circulation patterns, depicted in Figure 7.11.

Langmuir Circulation

A unique, small-scale circulation pattern occurs in a complex phenomenon called *Langmuir circulation*. Where winds blow steadily for long periods of time in the open ocean, alternating bands of convergence and divergence appear (Figure 7.12). The resultant upwelling and downwelling patterns establish small (perhaps 5 to 30 meters in diameter) circular cells between zones of convergence and divergence. Floating marine life and debris tend to become concentrated in downwelling regions and consequently arrayed in long parallel rows aligned with the wind direction. Langmuir circulation is particularly evident in the Sargasso Sea in the tropical North Atlantic, where it is occasionally manifested by straight, parallel lines of seaweed stretching as far as the eye can see. The phenomenon is not confined to the oceans and, under the right conditions, can even occur in small ponds.

FIGURE 7.12. Langmuir circulation. (After Weyl, P. K. 1970. *Oceanography.* N.Y.: Wiley.)

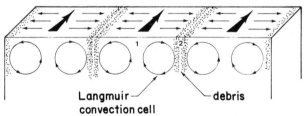

Langmuir convection cell — debris

 WIND DIRECTION

→ WATER TRANSPORT

1 UPWELLING (DIVERGENCE)

2 DOWNWELLING (CONVERGENCE)

A considerable portion of the oceans' circulation occurs unseen, far below the sea surface. While some of this deep-water movement is attributable to the wind-driven circulation, much of it is accounted for by the thermohaline circulation.

The thermohaline circulation owes its name to the fact that it is ultimately driven by the sinking of water masses, primarily in high latitudes, in response to temperature and salinity changes. Before describing the details of thermohaline circulation, let us first examine patterns of salinity, temperature, and density in the oceans.

Salinity, Temperature, and Density

The density of sea water, like that of any substance, is a measure of the water's weight or mass per unit volume and is usually indicated in units of grams per cubic centimeter (g/cm^3). Pure water at atmospheric pressure and room temperature has a density of $1.000 \ g/cm^3$. Primarily because of its salt content, sea water is slightly denser than fresh water. Ignoring pressure effects, its density usually lies between 1.024 and 1.028 g/cm^3. While a difference of only 0.004 g/cm^3 may seem like a narrow range, oceanographers readily calculate density to better than 0.00001 g/cm^3. Such precise measurements enable accurate and revealing determinations of the origin and history of a water mass.

The density of water is a function of its pressure, temperature, and salinity. The pressure effect results from reduction in volume attributable to compression. Since water is slightly compressible (a 1 percent reduction in volume occurs at a depth of 2 kilometers), the enhancing effect of pressure on density is important.

In addition to pressure, temperature and salinity also have significant effects on density. Either an increase in salinity or a decrease in temperature will result in an increase in density. Since the two factors have interacting effects on sea-water density, it is often difficult to precisely determine the independent effect of either on the density of a specific water mass. Oceanographers have historically made use of predetermined plots of the combinations of temperature and salinity, which yield lines of constant density. These plots are called *T-S diagrams* (Figure 7.13). The T-S diagram allows oceanographers to reconstruct the density profile of a column of water from profiles of salinity and temperature.

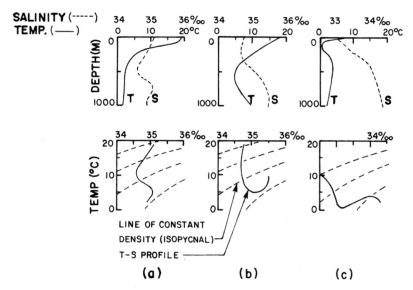

FIGURE 7.13. Hypothetical profiles of temperature and salinity with depth and corresponding T-S diagrams. (a) and (c) are possible but (b) is unlikely. (From Pickard, G. L., and Emery, W. J. 1982. *Descriptive physical oceanography.* 4th ed. Oxford: Pergamon Press.)

It is of interest to know how well the oceans are mixed. Correlations of temperature, salinity, and density with depth can yield this information. If the oceans were thoroughly mixed, we would expect density to be uniform from top to bottom.

Decades of physical sampling of the ocean depths have shown that the oceans are, in fact, poorly mixed. Indeed, they are highly *stratified.* This is indicated by vertical profiles of temperature and salinity which reveal layers of water masses with sharply defined boundaries. Why should this be so? Ultimately, the explanation lies in the fact that major changes in ocean density originate in the upper layers of the seas—the region which is exposed to the energy of the sun and the dynamics of the atmosphere. Let's have a closer look at these surface waters.

Density Changes in Surface Waters

The temperature and salinity of surface waters reflect the overlying atmospheric temperature and relative rates of evaporation and precipitation. The responses of surface waters to these factors are apparent on both a global and seasonal basis.

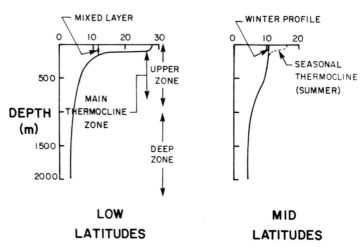

FIGURE 7.14. Characteristic profiles of temperature with depth in the open ocean. (From Pickard, G. L., and Emery, W. J. 1982. *Descriptive physical oceanography.* 4th ed. Oxford: Pergamon Press.)

Where air temperatures are warm year-round and evaporation rates are relatively low, as in the low latitudes, a layer of warm, low-density surface water overlies colder, denser water below. Between the two layers exists a zone of rapidly declining temperature (the *thermocline;* Figure 7.14) and rapidly increasing density (the *pycnocline*). Because the lighter water lies on top, such water columns are *stable;* that is, they are resistant to mixing. Year-round, stable water columns are characteristic of the tropics and are frequently deficient in marine life because the absence of mixing prevents necessary renewal of essential nutrients, trapped in deeper waters below.

Stable water columns also exist for part of the year in higher latitudes because of the existence of a *seasonal thermocline* (Figure 7.14). With the breakdown of the thermocline in late fall, the water column becomes unstable because, once cooled, the surface waters become denser than the waters below. Consequently, the surface waters sink, and the upper layers of the water column are mixed in the process. Seasonal mixing and renewal of nutrients in surface waters contribute to the comparatively high productivity of temperate-latitude marine environments.

Thermoclines and pycnoclines are comparatively short-lived and shallow in polar surface waters because there is no prolonged atmospheric heating of the upper layer of the oceans. Further, polar surface waters are the breeding grounds for vast quantities of the world's densest waters, which eventually sink to the depths. What is the origin of these dense waters? Mostly, it is cooling during high-latitude winters. The formation of sea ice is an important contributing factor.

The sinking of dense surface waters is most dramatically represented near Antarctica (Figure 7.15). There, ice formation in the Weddell Sea creates the densest ocean water in the world, with an average temperature of $-1°$ C and an average salinity of 34.6 parts per thousand. When sea water freezes, about 70 percent of the salt is excluded from the crystalline structure of ice. The salt accumulates below the ice, raising the salinity of the frigid liquid water there and

FIGURE 7.15. Surface water circulation of the Southern Ocean in the vicinity of Antarctica. (From Pickard, G. L., and Emery, W. J. 1982. *Descriptive physical oceanography.* 4th ed. Oxford: Pergamon Press.)

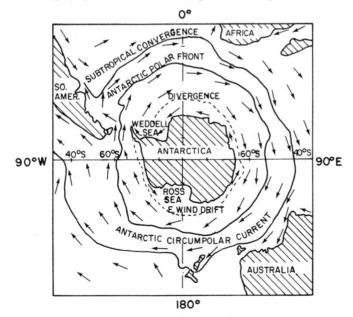

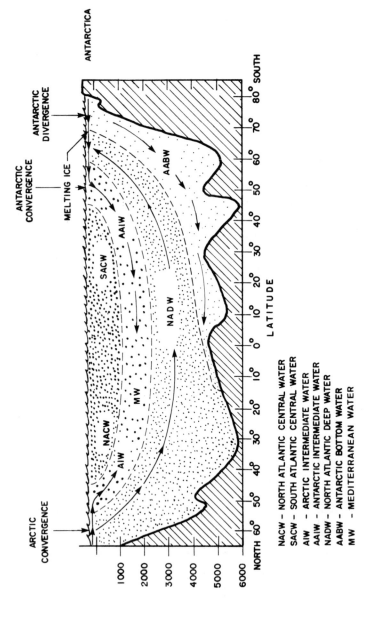

FIGURE 7.16. Generalized diagram of the subsurface circulation in the Atlantic Ocean. Boundaries between water masses are actually very diffuse. (After Anikouchine, W. A., and Sternberg, R. W. 1973. *The world ocean.* Englewood Cliffs, N.J.: Prentice-Hall.)

NACW – NORTH ATLANTIC CENTRAL WATER
SACW – SOUTH ATLANTIC CENTRAL WATER
AIW – ARCTIC INTERMEDIATE WATER
AAIW – ANTARCTIC INTERMEDIATE WATER
NADW – NORTH ATLANTIC DEEP WATER
AABW – ANTARCTIC BOTTOM WATER
MW – MEDITERRANEAN WATER

increasing the water's density. During ice formation in the Weddell Sea, the resultant cold, briny waters sink to the bottom, then flow toward the equator. This flow, called *Antarctic Bottom Water* (AABW), underlies much of the South Atlantic. Eventually, it is entrained or gradually mixes into other subsurface waters.

When it first sinks, AABW is rich in oxygen. There are several reasons for this. First, since the water mass originates at the surface, its dissolved gases equilibrate with those of the atmosphere. Second, cold waters can hold relatively large amounts of gases like oxygen. Finally, the well-mixed, nutrient-enriched waters around Antarctica stimulate high seasonal productivity, including large quantities of phytoplankton. These minute floating plants contribute significant quantities of oxygen to the water during photosynthesis.

A "parcel" of AABW may take hundreds of years to complete a journey from its point of origin in the Weddell Sea to the sea floor below the equator. Gradually, oxygen levels are reduced because there is no source of renewal once the water sinks below the surface and because of oxygen-demanding respiration and decomposition.

Winter ice formation also causes some sinking of high-latitude waters in the northern hemisphere; however, this *Arctic Bottom Water* (ABW) is largely restricted to the Arctic sea floor because of obstructing bottom topography.

The region which produces the greatest amount of deep water in the world oceans lies south of the Arctic, in the subpolar regions of the North Atlantic, east of Greenland. Cold temperatures and high salinity contribute to the formation and sinking of these large volumes of dense water, collectively referred to as *North Atlantic Deep Water* (NADW). The surface waters of the North Atlantic are among the saltiest in the world, averaging 35 parts per thousand, in part because the Gulf Stream delivers large volumes of saline waters to the region and in part because of relatively high evaporation rates. The waters are also cold, averaging from below zero to about 3° C. After sinking, these waters range far and wide. Antarctic Bottom Water may be a seasoned traveler, but North Atlantic Deep Water can truly be accused of wanderlust, coursing south and ultimately welling upward over a broad area of the South Atlantic (Figure 7.16). Dense as NADW is, it is slightly lighter than AABW. Where the two water masses converge, they segregate by density, with NADW overriding AABW.

The Antarctic region contributes still another major volume of water to the deep circulation. Cooling causes a major sinking of surface waters at about 50 degrees south latitude. Though cold (about

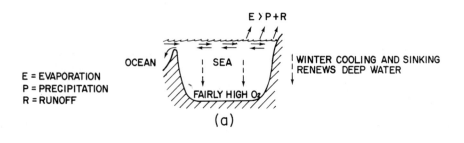

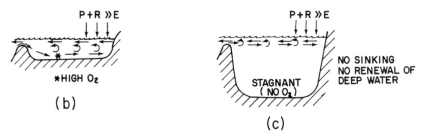

FIGURE 7.17. Circulation patterns in semienclosed seas. (a) Mediterranean Sea and Red Sea. Evaporation exceeds precipitation and runoff from land, increasing salinity in upper layers. In winter, cooled surface water sinks and renews deep water. (b) Baltic Sea and Hudson Bay. Precipitation and runoff greatly exceed evaporation. Shallow depth allows considerable vertical mixing. (c) Black Sea. Precipitation and runoff greatly exceed evaporation. A layer of low-salinity, low-density surface water creates a stable water column which, because of relatively great depth, is resistant to mixing. (From Pickard, G. L., and Emery, W. J. 1982. *Descriptive physical oceanography.* 4th ed. Oxford: Pergamon Press.)

$4° C$) and fairly salty (averaging 34.2 parts per thousand) this *Antarctic Intermediate Water* (AAIW) is warmer, less salty, and less dense than AABW and NADW, and consequently is confined to depths above those water masses, where it slowly flows toward the equator (Figure 7.16).

Some sinking of high-density waters also occurs in lower latitudes. Here, the elevated density is strictly a function of high salinity because of high evaporation rates. The waters of the Mediterranean provide an example. Because evaporation far exceeds precipitation in the virtually enclosed Mediterranean basin, the high salinity of surface waters causes continual sinking (Figure 7.17). Deep, saline (up to 39 parts per thousand) water flows out over the Gibraltar sill into the Atlantic Ocean, where is eventually reaches an equilibrium depth of about 1.2 kilometers (the depth at which Atlantic Ocean water has the same density). The Mediterranean water mass then spreads throughout the Atlantic at that depth, where it is detectable far from the Strait of Gibraltar.

Comparisons Between the Density-Driven and Wind-Driven Circulation

The density-driven circulation primarily involves the transport of water masses from cold, high latitudes to the remainder of the ocean basins. Eventually, these waters work their way to the surface again, primarily through slow, gradual mixing with overlying waters. Consequently, like the wind-driven circulation, the density-driven circulation accounts for a significant redistribution of global heat, helping to minimize temperature differences between the poles and the equator.

Density-driven currents do not course as swiftly through the ocean basins as their cousins, the surface currents. Typically, their velocities are 1 to 2 centimeters per second. Even along the western margins of ocean basins, where their speeds occasionally reach 40 centimeters per second, they flow less than 20 percent as quickly as the Gulf Stream.

Like wind-driven currents, thermohaline currents are clearly identifiable water masses, distinguishable from adjacent water masses because of unique characteristics of temperature, salinity, or dissolved gases. These characteristics can be used as *tracers* for the water mass, allowing oceanographers to track the path of a deep current far from its source. In contrast to surface currents, deep currents are segregated by density differences and not by wind patterns. In both cases, the currents are in part shaped by their interactions with land masses and/or bottom topography. Finally, the major oceanic gyres largely restrict wind-driven currents to individual ocean basins; there is relatively little transport of wind-driven waters across the equator. The density-driven circulation, on the other hand, accounts for considerable movement of waters between hemispheres.

SUGGESTED READINGS

BAKER, D. J., JR. 1970. Models of ocean circulation. *Scientific American* 221:114-21.

GROSS, M. G. 1982. *Oceanography: a view of the earth.* 3rd ed. Chapter 7. Englewood Cliffs, N.J.: Prentice-Hall.

JOYCE, T., AND WIEBE, P. 1983. Warm-core rings of the Gulf Stream. *Oceanus* 26:34-45.

MASON, P. 1975. The changeable ocean river. *Sea Frontiers* 21:171-77.

MUNK, W. 1955. The circulation of the oceans. *Scientific American* 191:96-108.

PICKARD, G. L., AND EMERY, W. J. 1982. *Descriptive physical oceanography.* 4th ed. Oxford: Pergamon.

STEWART, R. W. 1969. The atmosphere and the ocean. *Scientific American* 221:76-105.

SVERDRUP, H. U., JOHNSON, M. W., AND FLEMING, R. H. 1942. *The oceans: their physics, chemistry, and general biology.* Englewood Cliffs, N.J.: Prentice-Hall.

TCHERNIA, P. 1980. *Descriptive regional oceanography.* Oxford: Pergamon.

WIEBE, P. 1982. Rings of the Gulf Stream. *Scientific American* 246:60-70.

Chapter Eight

Ocean Circulation: Waves and Tides

A GROUP OF BIRD WATCHERS ROSE ONE RECENT WINTER MORNING and made their way to the frozen beach. As they moved toward the water's edge, their footsteps crunched quietly over ice-crystal sand castles. Subzero air banked against warmer ocean, and ragged puffs of sea smoke rolled and tumbled toward the horizon. The beach was quiet save for the intermittent screech of diving terns. All motion seemed slowed. A hundred yards offshore, eider ducks bobbed lazily on the congealed swells. At the sea's edge the waves themselves seemed reluctant to break, as if their shapes might shatter in the bitter, brittle dawn.

The birders recalled another February morning years past. A blizzard raged against the coast. Snow drove in horizontal sheets across the windswept sand, plastering its whiteness against inland trees. Beach and water were one as monster waves engulfed the shore.

Two winter days, two different scenes, but linked by the common phenomenon of ocean waves. Even on the calmest day of the year the sea still delivers its sinuous progeny to passive shores. Ocean waves are omnipresent, though it is a certainty that no two are alike. What are waves, and how are they engendered? What factors affect their height, their destructive power, their individuality, and their mutual interaction? How do they behave in shallow water and in the depths? The purpose of this chapter is to answer these questions and to impart some appreciation and understanding of one of the most ubiquitous but poorly understood phenomena in the oceans.

WAVES

Introduction and Terminology

The next time you see an ocean wave, watch it closely. Its shape may be imperfect, marred by surface ripples, a broken crest, or streaks of windblown foam, yet its sinusoidal form is clearly wavelike, resembling those classic forms in vaguely remembered physics texts. In fact, ocean waves obey the same fundamental laws of physics as sound and light waves. They are disturbances in a medium, generated by the transmission of energy. The disturbance consists of vibratory impulses which pass through the medium at a speed determined by the properties of the medium. In some situations the

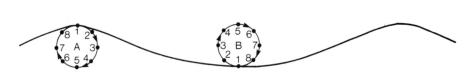

FIGURE 8.1. The simultaneous orbital paths (A, B) described by two water particles in a water wave. Numbers refer to the particles' sequential positions in time.

disturbance travels in a direction parallel to the direction of energy propagation. Such waves are *longitudinal waves.* Conversely, *transverse waves* move in a direction perpendicular to that of energy transmission.

Reexamine that ocean wave. Is there a piece of flotsam on its surface? Watch the object closely. It is not carried along in the direction of the wave but bobs up and down as the wave passes beneath it. An important property of waves is that, ideally, there is no net displacement of the particles set in motion by a wave. To visualize this more clearly, consider what happens when you snap a piece of rope held in your hand. A wave moves along the rope, but the individual fibers do not. This is not to say that the fibers do not move at all. In fact, they move up and down. Similarly, if we were able to follow the path of a single water molecule set in motion by an ocean wave, we would see that it undergoes essentially no forward displacement. Instead, it describes a circular orbit (Figure 8.1). *Orbital motion* of particles is characteristic of waves which pass between fluids of different densities, such as at an air-water interface. (In contrast, waves generated in solid media cause particles to move back and forth in a longitudinal or transverse direction.)

In reality, water molecules affected by an ocean wave undergo a very slight net forward displacement because of interacting forces of gravity and inertia. Were waves to cause significant forward displacement of water, ships would routinely be subjected to enormous stresses. *Breaking waves* do, however, transport water forward, with resultant damage to fixed and floating structures.

With this background, let us introduce some terminology specific to wave phenomena. Consider the idealized wave depicted in

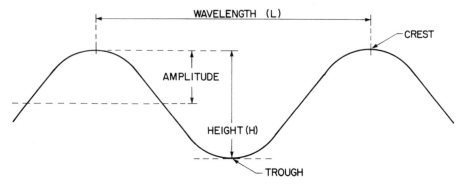

FIGURE 8.2. Wave terminology.

Figure 8.2. The *crest* of a wave is its highest point, and the *trough* represents its lowest. *Wavelength* (L) is the distance between successive crests or troughs. *Wave height* (H) is the vertical distance between trough and crest, while the *amplitude* is one half the height. *Wave period* (T) is the time for two successive crests or troughs to pass a fixed point of reference. *Frequency* (f) refers to the number of wavelengths which pass a fixed point in a given time.

The frequency ($1/T$), *velocity* (L/T), and *steepness* (H/L) of waves can be calculated if wavelength, height, and period are known (Figure 8.2). For example, consider a series of waves which passes by your vantage point on a pier. By carefully timing the passage of waves and estimating their height and wavelength by comparisons with reference points on the pier, you estimate that their average wavelength is 100 meters ($L = 100$m), their height is two meters ($H = 2$m), and that it takes 10 seconds for one wavelength to pass ($T = 10$s). Thus, their frequency is one wavelength every 10 seconds ($1/10$s), or 6L per minute (6L/min.). The average wave velocity is 100m/10s or 10 meters per second (10m/s). The steepness of the waves is 2m/100m or 1/50. Since waves usually do not break until the steepness exceeds 1/7, the gentle, undulating nature of these waves is not surprising.

Wind Waves in the Ocean

Ocean waves can be caused by a variety of phenomena. Primary among these are the wind, seismic disturbances such as earthquakes and underwater volcanic eruptions, and the gravitational attraction between the earth and the sun and the moon which causes the long-period and long-wavelength waves known as the *tides*. Let us first consider wind-generated waves, perhaps the most obvious wave forms in the ocean.

Imagine a preposterous proposition—a completely smooth sea surface. Observe its response to a developing storm. As the wind begins to blow, the slight disturbance of the sea surface will first be manifested as tiny round ripples or *capillary waves* (Figure 8.3). Capillary waves have extremely small wavelengths (less than 1.74 centimeters), short periods (less than 0.1 second), and sufficiently small height that the dominant restoring force which "seeks" to dampen them and return the sea surface to its glassy state is surface tension. This restoring force propagates the wave in a horizontal direction, much as a ripple travels along a tightly stretched piece of fabric.

As the wind continues to blow, waves grow in length, height, and energy, since the disturbed sea surface is increasingly more directly exposed to the wind. While we still have a lot to learn about the effect of wind on wave formation and behavior, some responses have been relatively well defined. At lengths greater than 1.74 centimeters, gravity replaces surface tension as the dominant restoring force, and the resultant form is called a *gravity wave*. With the formation of gravity waves, a short, choppy sea characteristically develops, since the interactions between different waves create a variety of wave forms with different wavelengths (Figure 8.4). As more and more energy is imparted to the waves, wave height increases more rapidly than wavelength until, eventually, the resultant steepness causes the waves to break. In breaking waves, or *whitecaps*, the energy received from the wind is balanced by the energy lost in breaking.

FIGURE 8.3. The development of waves on an initially smooth sea surface in response to a sustained unidirectional wind (indicated by arrows).

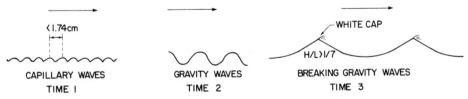

<1.74cm

WHITE CAP

H/L>1/7

CAPILLARY WAVES
TIME 1

GRAVITY WAVES
TIME 2

BREAKING GRAVITY WAVES
TIME 3

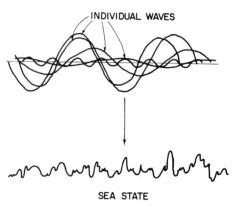

FIGURE 8.4. Interactions between individual waves of different lengths and heights create a choppy, confused sea state.

SWELLS, WAVE GROUPS, AND WAVE TRAINS. The choppy sea which develops at the wind's source consists of a myriad of wave forms with different lengths, heights, and periods. The interaction, or *superposition*, of different waves further modifies the original wave forms and increases the sea's confusion. However, observe the waves further from the storm's origin (Figure 8.5). The choppy sea state is gone, replaced by battalions of waves of uniform sizes and lengths, called *swells*, precision-marching to a distant shore. What turned confusion into order? To understand what has happened we have to know something about the factors which affect the speed of waves. In deep water, where the ocean bottom is too far away to influence the progress of a wave, the wave's speed is primarily a function of its wavelength and period. In essence, the longer the period or the wavelength, the faster the wave. Consider, then, the response of our myriad wave forms in the confused seas near the storm's origin. As the waves move away from the source of the disturbance, the longest ones move at the fastest speed. In this process of "sorting themselves out" by wavelength, the swells segregate themselves into *wave groups* (Figure 8.5) corresponding to their speed and wavelength. This principle of wave *dispersion* means that the first waves to break on a distant shore will be the longest, fastest waves. These will be followed by sequences of groups with progressively shorter, slower wavelengths.

Now watch a single wave group closely. The lead wave gradually disappears, only to be replaced by the wave behind it. Soon that wave too vanishes, with its place also taken by its follower. At the

WIND AND WAVE DIRECTION

O = ORIGIN OF STORM ; CONFUSED, CHOPPY SEA

I = WAVE GROUP I

2 = WAVE GROUP 2 (with longer wavelength and greater velocity,
    it has "outrun" WAVE GROUP I)

FIGURE 8.5. The principle of wave dispersion. Wave groups ''sort themselves out'' as they move away from the origin of a storm, since the speed of a wave is a direct function of its wavelength.

rear of the wave group, new waves spring up as lead waves are lost, preserving the number and form of waves in the group. What is happening? The lead wave in a group expends considerable energy in disturbing the relatively quiet water before it. With its energy dissipated, it soon dies away. The energy "lost" at the front of the group is replaced by energy "gained" at the rear of the group, which, in turn, causes the creation of a new wave there. This phenomenon has the interesting result that the speed at which an individual wave moves in a group is twice that of the wave group itself. While difficult to conceptualize, the process is depicted in Figure 8.6.

FIGURE 8.6. The velocity of individual deep-water waves in a group is twice that of the wave group. (From Stowe, K. 1983. *Ocean science.* 2nd ed. N.Y.: Wiley.)

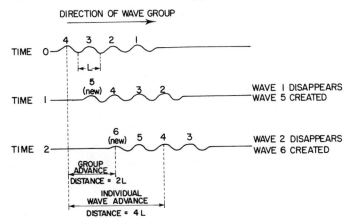

WAVE HEIGHT AND ENERGY. Many of us have read or heard stories about monster waves from sailors returned from harrowing encounters at sea. Almost without exception, such stories are to be taken with a grain of salt. Sailors' accounts of storms and waves are perhaps second only to fishermen's tales in their degree of embellishment and exaggeration. Occasionally, however, these reports are authenticated. In February 1933, a U.S. Naval officer aboard the Navy tanker U.S.S. *Ramapo* in the North Pacific accurately measured a 34-meter wave. As the wave towered behind the ship, the young lieutenant coolly sighted through a boom on the crow's-nest, aft of his vantage point on the bridge, to the wave crest astern of the vessel. When the boom and the crest were in line, the horizon behind the wave was obscured. Some simple calculations involving the length of the ship and the height of the crow's-nest and bridge above the ship's waterline allowed the sailor to estimate the wave's height. The officer's considerable presence of mind is worth noting—a 34-meter wave is about the height of a 10-story building! No higher deep-water wave has ever been authenticated. Waves in shallow water do frequently grow to greater heights, but the factors affecting their heights are different from those influencing deep-water waves. Shallow-water waves will be discussed later in this chapter.

Wind imparts energy to waves which is manifested by increases in wave height, length, velocity, and period. In examining the effects of wind on waves, three factors must be considered. These are (1) wind velocity, (2) how long the wind has blown from one direction (*duration*), and (3) the unobstructed distance over which the wind has blown from the same direction (*fetch*).

At a specific wind speed, there is a maximum wave height possible, even with unlimited fetch and duration of blowing. Similarly, limited fetch and/or duration will limit wave height, even in the strongest gales. The largest waves possible for a given wind speed, when fetch and duration are maximal, constitute a *fully developed sea.*

WAVE INTERFERENCE. Even in deep water, accurate knowledge of wind speed, duration, and fetch seldom allow an accurate prediction of wave size and behavior. For one thing, inevitably more than one storm will be contributing to the waves in any given area. Wave trains approaching from different directions will interfere with each other at their point of convergence. The resultant interference pattern of two converging waves will be the sum of the individual disturbances of each wave at the point of intersection. What this means is that if two waves come together exactly in phase, a new wave will be created whose height is the sum of the heights of the two individual waves. This is known as a *constructive interference* pattern. Conversely, two waves approaching out of phase will produce a wave whose height is less than that of the higher wave. The extreme case of this *destructive interference* occurs when two waves of exactly equal heights and lengths come together from opposite directions, 180° out of phase. The result will be a completely smooth sea surface. Normally, wave trains from several ocean disturbances interfere in a confused fashion at any location in the ocean, producing a *mixed interference* pattern in which some waves interfere constructively and some destructively.

The unpredictable interference patterns established by wave trains converging from different directions can have catastrophic results. On August 14, 1979, a series of enormous "rogue" waves battered a 303-boat sailing fleet participating in the Fastnet Race off the southwest coast of Great Britain. Twenty-three boats were lost, and fifteen people died in the ensuing havoc. Later reconstruction of the meteorological events which spawned the giant waves strongly implicated constructive interference as a major contributing factor. A single storm moving through the area produced wave trains from two different directions as it changed course. Though the winds had been blowing only about 18 hours at the time the monster waves struck, the pattern resulting from their convergence produced waves 15 meters high.

WAVE REFLECTION AND DIFFRACTION. The behavior and form of waves, including wind-generated waves, in deep water is also modified when they strike solid objects. When an entire wave rolls against a barrier, such as a sea wall, the wave is reflected by the barrier. When only part of a wave strikes a barrier, the unobstructed portion of the wave bends around or is diffracted by the barrier. Let's examine each of these situations in more detail.

Wave Reflection. Watch a wave closely as it surges against a sea wall. Obviously, since the wall blocks the forward progress of the wave, its orbital motion is interrupted. Simultaneously, as the wave is reflected by the barrier, a new wave is generated in the opposite direction with a reciprocal orbital motion (Figure 8.7). The resultant meeting of the two waves produces an interference pattern

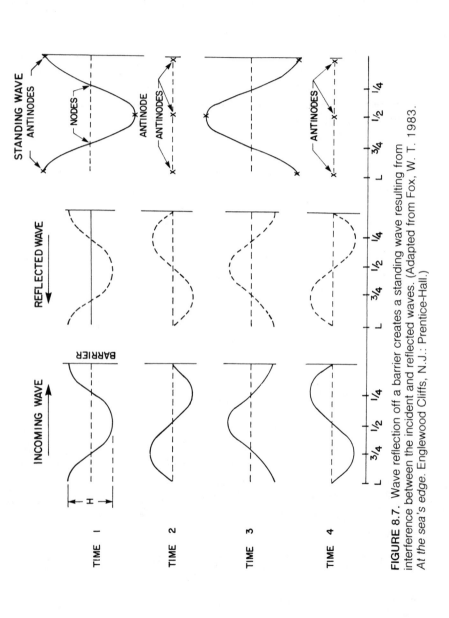

FIGURE 8.7. Wave reflection off a barrier creates a standing wave resulting from interference between the incident and reflected waves. (Adapted from Fox, W. T. 1983. *At the sea's edge.* Englewood Cliffs, N.J.: Prentice-Hall.)

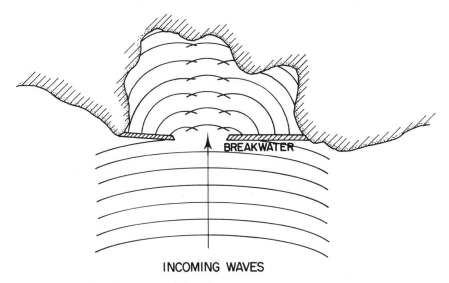

INCOMING WAVES

FIGURE 8.8. Wave diffraction. (After Stowe, K. 1983. *Ocean science*. 2nd ed. N.Y.: Wiley.)

which may be constructive, if the crests of the two waves are in phase (i.e., wave crests are superimposed), or destructive, if they are out of phase (i.e., crests and troughs are superimposed), or mixed (Figure 8.7). The interaction of the incoming and reflected waves thus sets up a *standing wave* which is characterized by an oscillatory pattern of the sea surface. At specific fixed points, called *nodes*, there is no vertical displacement of the water. On either side of the nodal points, the water oscillates up and down. Maximum vertical displacement of the water surface occurs at points midway between nodes called *antinodes*.

Wave Diffraction. Every point on a wave is a source of energy capable, in itself, of generating a ring of tiny wavelets radiating out from the point source in all directions. An apt analogy would be the circular ring of wavelets initiated by a stone dropped vertically into a quiet pond. Because wind-generated waves tend to be unidirectional, interference between adjacent wavelets eliminates all but the leading edges of their radial arcs. The result is a single, relatively high unidirectional wave which moves in the same direction as the wind and is, in a sense, a summation of the individual wavelets. When such an ocean wave meets a partial barrier, like a breakwater, a peculiar result occurs. The part of the wave which is not blocked may actually bend around the breakwater, focusing wave energy on the very anchorage which was supposed to be protected by the breakwater (Figure 8.8). This bending pattern, or *wave diffraction*, results because when part of a wave is blocked by a barrier, the remainder of the wave is free to generate radial wave energy without interference on the side nearest the barrier.

Before discussing the behavior of waves in shallow water, it is important to point out that wave reflection and diffraction occur in shallow water as well as in deep water, though the proximity of the ocean bottom may modify the response. Further, waves generated by forces other than the wind may undergo reflection and diffraction.

THE BEHAVIOR OF WAVES IN SHALLOW WATER. It has long been recognized that when waves move into sufficiently shallow water, they behave rather oddly. In fact, the influence of the bottom on the form and behavior of waves is so pronounced that it has become necessary to study and describe shallow-water waves as though they were a species distinct from those forms which are found only in deep water. This is a matter of convenience rather than fact, since waves may actually be considered as constituting a spectrum between extremely shallow-water and extremely deep-water forms.

Scientists have arbitrarily defined *shallow-water waves* as waves whose length is at least 20 times the depth of the water (i.e., $L \geqslant 20Z$ or $Z \leqslant 1/20L$, where $L =$ wavelength and $Z =$ depth) and *deep-water waves* as those whose length is less than or equal to twice the water depth (i.e., $L \leqslant 2Z$ or $Z \geqslant 1/2L$). In other words, for shallow-water waves, wavelength is large relative to water depth, while for deep-water waves, water depth is large compared to wavelength. Shallow-water waves are said to "feel the bottom," which essentially means that their contact with the bottom affects their behavior.

To understand how waves in shallow water are affected by the bottom, consider the circular orbits of wave particles. In deep water, the circular orbits become smaller with depth but retain their round shapes (Figure 8.9). At depths greater than $1/2L$, there is little or no water movement. In contrast, in shallow water the orbital motion of particles becomes increasingly flattened, or elliptical, with depth (Figure 8.9). In fact, just above the bottom the water describes a horizontal, back-and-forth motion. You have probably observed this if, while skin diving in shallow water, you have noticed the rhythmic sway of sea grasses responding to the beat of waves overhead. In essence, the proximity of the bottom is deforming the orbits of shal-

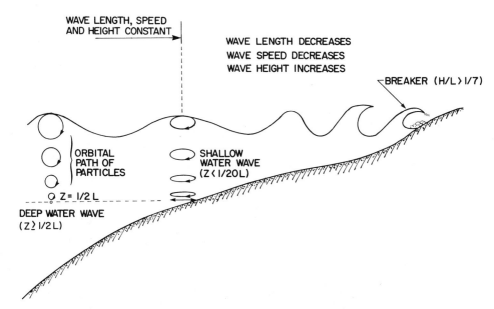

WAVE LENGTH, SPEED
AND HEIGHT CONSTANT

WAVE LENGTH DECREASES
WAVE SPEED DECREASES
WAVE HEIGHT INCREASES

BREAKER (H/L > 1/7)

ORBITAL
PATH OF
PARTICLES

SHALLOW
WATER WAVE
(Z < 1/20L)

Z = 1/2 L

DEEP WATER WAVE
(Z ≥ 1/2 L)

FIGURE 8.9. Deep- and shallow-water waves. (Adapted from Anikouchine, W. A., and Sternberg, R. W. 1973. *The world ocean*. Englewood Cliffs, N.J.: Prentice-Hall.)

low-water waves, restricting their vertical displacement without confining their horizontal movement.

The proximity of the bottom also influences the velocity of shallow-water waves. Thus, calculations of wave speed in water depths of less than 1/20 of the wavelength must include a term for water depth, while the speed of deep-water waves is a function of wavelength or period but is independent of water depth.

Most of us have stood on a beach and watched a steady parade of waves marching in disciplined close-order drill toward the shore. At the limit of visibility the waves are long, low shapes snaking slowly toward the coast, but near shore these placid forms turn hostile, rearing up and dashing themselves against the sand in foaming, profligate abandon. What is responsible for this ornery behavior? Let's isolate and examine a single wave to understand what is going on.

Once our wave begins to make contact with the bottom, it slows down. The reduction of velocity compresses the wavelength without proportionately reducing the wave's energy. In addition, the reduction in velocity allows the following wave to catch up. Consequently, the height of the wave increases significantly (Figure 8.9). Once the wave gets high enough (steepness, or H/L, exceeds

SPILLING BREAKER   PLUNGING BREAKER   SURGING BREAKER

FIGURE 8.10. Kinds of breakers. The type of breaker occurring on a beach is a function of wave steepness and bottom topography. (From Gross, M. G. 1982. *Oceanography.* 3rd ed. Englewood Cliffs, N.J.: Prentice-Hall.)

1/7), it will break, dissipating energy in the process. Breaking waves in shallow water are called *breakers;* the various kinds of breakers, in totality, are called *surf.*

BREAKERS AND SURF. An experienced surfer recognizes several varieties of breakers (Figure 8.10). *Spilling breakers* occur on relatively flat beach slopes where the waves rise slowly and gradually. Once they become steep enough to break, the gentle bottom slope allows them to dissipate their breaking energy over a large expanse of the surf zone. Thus, the breaking water gradually spills over the face of the wave, resulting in a long-lasting wave particularly favored by surfers. The spectacular *plunging breakers,* with their characteristic curling crest, large air pocket, and mass of foam preceding the wave form offer an especially exciting ride, since their energy is highly concentrated in the breaking region. They occur when the beach slope is slightly steeper, ranging from three to eleven degrees. On even steeper beach slopes, *surging breakers* characteristically occur. Here, the long, low waves do not have a chance to break before meeting the steep beach face. In fact, the wave is often reflected off the beach, resulting in the creation of standing waves near shore.

WAVE REFRACTION. Have you ever stood on a beach and wondered why the waves always come in nearly parallel to the shore, even when the waves offshore are arrayed at an obvious angle to the coastline? You have witnessed still another phenomenon attributable to the behavior of waves in shallow water. Recall the reduction in velocity which occurs when a wave begins to "feel the bottom." As a wave approaches the shore at an angle, that portion of the wave which reaches shallow water first will be the first to slow down. The rest of the wave then progressively "catches up" to the slowed portions of the wave nearer shore until eventually the entire wave is affected by the bottom (Figure 8.11). This causes the wave to "bend" toward shore, with the result that it approaches the beach in a parallel orientation. This bending of waves in shallow water because of contact with the bottom is called *refraction*. The principle of wave refraction is responsible for the surprising observation that ocean waves break all the way around the perimeter of an island despite the direction of waves offshore (Figure 8.12).

Refraction also accounts for the fact that coastal promontories and headlands are particularly vulnerable to wave damage and destruction (Figure 8.13). Be thankful you are not a lighthouse keeper. Even on relatively calm days, your location on a rocky point will treat you to views of crashing eruptions of water, while your friends in an adjacent cove are lolling around in comparatively placid waters. On stormy days the results can be intimidating for even the stoutest hearts. There have been many hair-raising tales of large boulders being hurled against the tops of lighthouses by enormous waves boiling against coastal headlands. Unusual bottom topography can amplify this violence. A well-publicized case occurred in April 1930 in Long Beach, California. On a calm day a series of large breakers destroyed the end of the harbor breakwater by dislodging stones weighing up to 20 tons. The breakwater had withstood the ravages of major storms before. How could such destructive waves occur in the absence of a storm? The explanation

FIGURE 8.11. Waves approaching a shore undergo refraction causing them to break parallel to the shore. (From Stowe, K. 1983. *Ocean science.* 2nd ed. N.Y.: Wiley.)

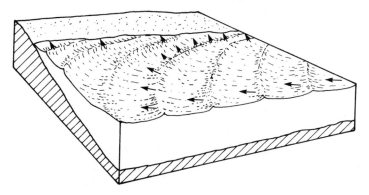

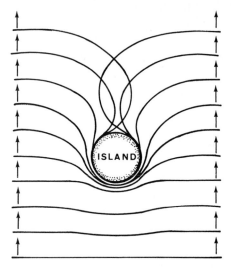

FIGURE 8.12. Refraction causes waves to break on the shore all around a small island, despite the direction of the wave train approaching the island. (From Stowe, K. 1983. *Ocean science.* 2nd ed. N.Y.: Wiley.)

lay in an anomaly in the underwater topography, 16 kilometers south of the breakwater. An unusually shallow area there caused refraction of incoming waves so that their energy was focused on the breakwater. Consequently, offshore waves about a half meter high were converted into waves with a 4-meter height at the breakwater.

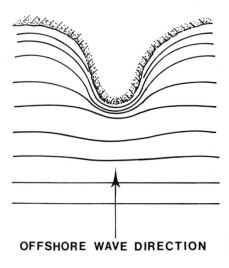

OFFSHORE WAVE DIRECTION

FIGURE 8.13. Refraction causes wave energy to be concentrated on headlands.

Not all ocean waves are generated by the wind. The most destructive waves on earth have nothing to do with the winds at all. These are *seismic sea waves*, popularly referred to as "tidal waves." In fact, the term *tidal wave* is inappropriate for these waves, since a true tidal wave—the tide itself—responds to gravitational and centrifugal interactions between the earth and the moon and the sun. In contrast, seismic sea waves result from geological activity in the earth's crust. This activity is characteristically "seismic"; that is, it results from earth-vibrating forces, such as earthquakes, volcanoes, or underwater landslides.

To avoid the confusion between tidal waves and tides, many oceanographers also use the Japanese name *tsunami* to describe seismic sea waves. Although the word has a nice sound to it and is probably easier to say than "seismic sea wave," it turns out that a literal English translation of tsunami is "tidal wave."

Let's examine the formation and behavior of a seismic sea wave. A major underwater seismic disturbance displaces a large volume of water from the origin of the disturbance which, in turn, causes a sudden change in the level of the sea surface above. A wave is then generated which moves radially outward from the disturbed area. The Pacific Ocean is particularly susceptible to seismic sea waves because of the large amount of tectonic activity in and around the Pacific Basin (see Chapter 4).

Seismic sea waves have unusually long wavelengths—typically more than 200 kilometers. Since the average ocean depth is about 4000 meters, wavelengths are about 50 times water depth, categorizing these waves as shallow-water waves whose velocity is a function of water depth. Because of their long wavelength seismic sea waves move rapidly—usually more than 700 kilometers per hour. At this speed a wave originating off the coast of California would reach the Asian mainland in less than a day.

Since most of the energy of a seismic sea wave is incorporated in its extremely long wavelength, their height in the open ocean seldom exceeds 0.5 meter. Monstrous, deep-water seismic sea waves, such as that depicted so sensationally in the movie *The Poseidon Adventure*, simply do not exist. How then do they get their well-deserved reputation for awesome destructive power? Once a seismic sea wave approaches shallow waters near shore, its velocity slows appreciably and its wavelength is significantly reduced. The resultant increase in height, characteristic of any wave entering shallow water, is especially dramatic for seismic sea waves because an enormous amount of energy is stored in its unusually long wavelength. It is not uncommon for a 30-meter-high wave to batter a coastline in the wake of a seismic disturbance.

Perhaps the most catastrophic seismic sea wave ever recorded was generated when the Indonesian island, Krakatoa, blew up in an immense volcanic eruption on August 27, 1883. The dust blown into the atmosphere had significant effects on the earth's weather for the next two years, but the most immediate result was a huge wave which pummeled adjacent coastal regions, killing 36,000 persons. The wave was ultimately detected in the English Channel, halfway around the world, which gives some sense of the enormous energy it contained.

Until fairly recently, there was little warning of an approaching seismic sea wave, since its triggering force might have been an event on the other side of an ocean. About the only notice given was a sudden lowering of water level along a coastline which corresponded to the wave trough which preceded the destructive crest. Often, the recession of coastal waters had the opposite of desired effects—instead of encouraging flight to higher ground, it often lured curious persons to exposed flats where they were inundated only minutes later.

After an extremely destructive wave struck Hawaii in 1946, there was a redoubling of efforts to predict seismic sea waves well in advance of their arrival. An International Tsunami Warning System was initiated in 1948. It works on the following principle: By precisely monitoring a number of strategically located tide-measuring devices, the anomalous displacement of the sea surface which signifies an underwater seismic disturbance can be detected. Determination of the pattern of such displacements enables oceanographers to predict the path and time of arrival of a wave. While the system works well for locations some distance from the seismic disturbance, it is of little use to areas near the wave's origin.

STORM SURGES

Some of the instances of greatest destruction of coastal property are caused not by true waves but by the low atmospheric pressure associated with storms. In the vicinity of a storm, the reduced pressure

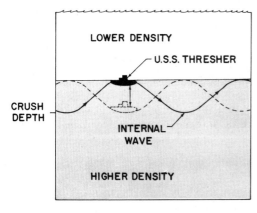

FIGURE 8.14. Hypothetical scenario depicting the possible role of an internal wave in the loss of the U.S.S. *Thresher* in 1963.

of the atmosphere on the sea surface allows the water to hump up into a domelike elevation. When the storm approaches a coast, the elevated sea surface causes unusually high water, particularly if the arrival of the storm coincides with the normal time of high tide. The effects of these *storm surges* (sometimes also referred to as *storm tides*) can be disastrous. In 1900 a storm surge flooded low-lying Galveston, Texas, causing 2000 deaths. A series of storm surges in the Bay of Bengal, near India, drowned an estimated half million people in 1970.

INTERNAL WAVES

Waves which occur completely underwater are known as *internal waves*. Otherwise behaving like waves at the interface of air and water, internal waves travel along the boundary between water masses of different densities in the ocean. These waves are not well understood, but it is believed that they can be caused by variable winds at the sea surface, by tidal forces, by seismic disturbances, or even by ships' propellers. Far less energy is required to initiate internal waves than surface waves. Their periods—measured in minutes to hours or days—are much longer than those of surface waves, and their speeds are much slower. They apparently can achieve formidable heights—perhaps exceeding 100 meters.

On April 10, 1963, the nuclear-powered submarine U.S.S. *Thresher* went down with all hands off the coast of Massachusetts. The loss of the submarine and its 129 crew members has never been adequately explained, but some researchers believe that internal waves may have been responsible. There is some evidence that the submarine was near its maximum operating depth at the time of its disappearance. If so, the vertical displacement caused by an internal wave could have been enough to bring the boat down to crush depth (Figure 8.14).

The tides are familiar to all of us, but few people realize that they are waves which display the characteristic properties of all orbital waves. Like seismic sea waves, tides are shallow-water waves with extremely long wavelengths—approximately half the circumference of the earth—but, unlike seismic sea waves, their arrival is predictable, regular, and seldom accompanied by destruction.

There are a number of interrelated forces causing and affecting the tides. Tidal behavior is complex, but tides have been studied for centuries. In fact, Sir Isaac Newton worked out the basic theory for the mechanics of the tides in the seventeenth century (see Chapter 1). Let's examine the major forces influencing tidal behavior on earth.

Gravitational and Centrifugal Forces

Ultimately, the tides are caused by the gravitational attraction between the earth and other celestial bodies, specifically the moon and the sun. While the apparent influence of those bodies on the tides of the earth has been described since Pliny the Elder studied them in the first century A.D., not until Sir Isaac Newton's work in the seventeenth century were the mathematical relationships elucidated.

According to Newton's universal law of gravitation, every object in the universe is attracted to (and attracts in turn) every other

FIGURE 8.15. The orbit of the earth-moon system around the sun. While the moon appears to orbit the earth, in fact the earth and the moon together move around a common center of gravity.

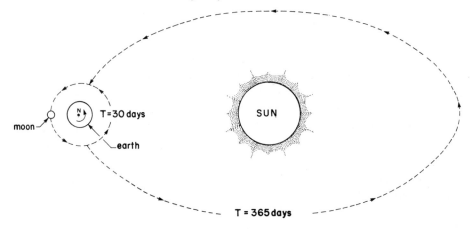

object in the universe with a force that is directly proportional to the product of their masses and inversely proportional to the square of the distance between them. The mathematical expression of the law is:

$$\text{Gravitational force} = G \frac{m_1 m_2}{r^2}$$

where G represents the universal gravitational constant, m_1 and m_2 are the masses of the respective objects, and r is the distance between the centers of the objects. Because of their relative proximity and size, respectively, the moon and the sun are the only celestial bodies which experience significant gravitational interactions with earth.

To simplify our initial discussion let us first consider only the interaction between the moon and the earth. Examine first the behavior of the earth and moon in space. While the moon revolves around the earth with a period of 30 days, the earth-moon system revolves around the sun with a period of slightly more than 365 days (Figure 8.15). Because the earth is 82 times as massive as the moon, this system is not evenly balanced: Its center of gravity does not coincide with the geographical center of the two bodies but instead lies at a point within the earth itself (Figure 8.16). As a consequence, as the earth-moon system moves around the sun, it behaves like a lopsided dumbbell, wobbling slightly about its center of gravity. This also means that the moon does not orbit a stationary body but, rather, that both earth and moon move around a common center called the *barycenter*.

Now consider a single point on earth. As the earth moves around its barycenter, that point will move in a circular orbit whose radius corresponds to the distance of the barycenter from the center of the earth (Figure 8.16). Gravitational forces cause every particle on earth (including particles of water) to be attracted to the moon, but the force of the attraction varies with the distance of the particle from the moon (Figure 8.17). Were there not a counteracting force, particles on earth would accelerate toward the moon (and vice versa) at a rate proportional to the gravitational force between them (since force equals mass times acceleration). Fortunately, *centrifugal force* is the counteracting force which prevents a collision.

Centrifugal force arises when a body moves in a curved path. It is exerted in a direction outward from the center of the circle described by the arc of the curved path. Examples abound. Have you ever rounded a corner too quickly in your car? If you turned to the right, you undoubtedly felt yourself being thrown against the driver's door. Centrifugal force was propelling you outward from the center of the radial arc. In much the same manner, the rotation of the earth around the barycenter tends to propel each particle on earth outward from the center of the particle's orbit. This propelling centrifugal force is equal for every particle on earth and is exerted in a direction parallel but opposite to a line connecting the centers of the earth and moon (Figure 8.17). Just as centrifugal force prevents the earth and moon from crashing together, gravitational attraction

FIGURE 8.16. The earth and the moon move as a system about a common center of gravity (the barycenter). Because the earth is much heavier than the moon, the barycenter is located much closer to the center of the earth than to the center of the moon (in fact, the barycenter is located within the earth itself). Consequently, the system behaves like a lopsided dumbbell. Each particle on earth moves in a circular orbit whose radius corresponds to the distance of the barycenter from the center of the earth. (Adapted from Fox, W. T. 1983. *At the sea's edge.* Englewood Cliffs, N.J.: Prentice-Hall.)

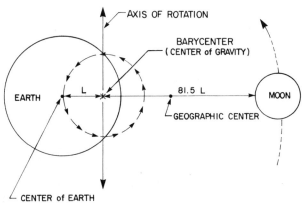

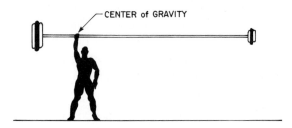

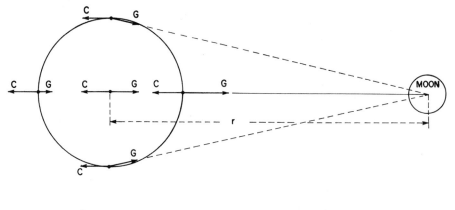

C
← CENTRIFUGAL FORCE VECTOR (ALL SAME LENGTH AND DIRECTION)

G
→ GRAVITATIONAL FORCE VECTOR (LENGTH AND DIRECTION VARY)

FIGURE 8.17. Depiction of the gravitational and centrifugal force vectors that establish the tide-producing force with the moon located over the earth's equator. For simplicity, gravitational force is shown here as occurring between the earth and the moon only, but there is also a significant gravitational attraction between the earth and the sun. (Adapted from Sverdrup, H. U., Johnson, M. W., and Fleming, R. H. 1942. *The oceans*. Englewood Cliffs, N.J.: Prentice-Hall; and Weiss, H. M., and Dorsey, M. W. 1979. *Investigating the marine environment: a sourcebook*. Groton, Conn.: Project Oceanology.)

keeps them from flying apart. The balance between the two forces keeps them in their characteristic orbits.

The Tide-generating Force

Let us now examine how the interaction of gravitational force and centrifugal force produces the *tide-generating force*. You will remember that the magnitude of the gravitational attraction of particles on earth to the moon varies with the distance of the particle from the moon. Since the gravitational attraction is along a line connecting the particle to the center of the moon, its direction also varies according to the location of the particle on earth (Figure 8.17). In contrast, recall that the centrifugal force is equal in magnitude and direction everywhere on earth.

A force with a magnitude and direction can be represented by an arrow called a *vector*. The length of the vector corresponds to the magnitude of the force, and its direction corresponds to the direction in which the force is exerted. Two interacting forces exert a net magnitude and direction which can be determined by comparing their individual magnitudes and directions.

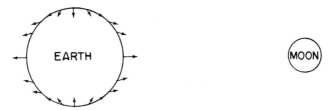

FIGURE 8.18. Arrows indicate the net tide-producing forces for various points on the earth with the moon located over the earth's equator. (From Sverdrup, H. U., Johnson, M. W., and Fleming, R. H. 1942. *The oceans*. Englewood Cliffs, N.J.: Prentice-Hall.)

By applying vector principles, we can calculate the *net tidal force* for every point on earth (considering only the effects of the moon for now). This is shown in Figure 8.18. The diagram reveals a surprising result: The net tidal forces are, on our idealized world, exactly equal in magnitude and opposite in direction on opposite sides of the earth. Since the net tidal force is the force exerted on the earth's waters, wherever the arrows point outward from the surface of the earth a tidal bulge which represents a high tide will occur. The relative height of the tide at any location will correspond to the vertical distance between the surface of the earth and the tip of the arrow. Low tides occur where arrows point in toward the earth.

In considering the behavior of the tides at any location on earth, don't forget that the earth rotates on its axis with a period of 24 hours. This means that, since tidal bulges occur simultaneously on opposite sides of the earth, most points on earth will experience two high tides and two low tides a day. However, the tidal cycle is complicated by the movement and position of the moon relative to a location on the earth. Let's look at this a little more closely.

Tidal Days and Tidal Periods

If you have ever spent any time at the shore, you have probably realized that the time of high tide is about an hour later each day. The prediction of tides at any location on earth must consider that a *tidal day* is more than 24 hours long. How can this be?

Imagine that you are viewing the earth from outer space and that you are looking down directly over the north pole (Figure 8.19). Further assume that the moon is directly over the equator.

What happens as the earth rotates on its axis? At time 0, a person standing on the equator will experience a fully high tide. After 6 hours the tide will be low, after 12 hours high again, and after 18 hours the tide will again be low. After 24 hours you would expect another fully high tide. However, you are forgetting something. Even as the earth rotates on its axis, the moon is orbiting the earth. Though it takes 30 days for the moon to go completely around the earth, after one day it will have traveled about 12 degrees (360/30), or 1/30 of the way. Therefore, it will take about 1/30 of a day (about 50 minutes) for a person on the equator to "catch up" to the new position of the moon and be exactly under the maximum bulge again, corresponding to a fully high tide. Thus, a *tidal day*, defined as the time interval between two successive transits of the moon over a local meridian (line of longitude), is 24 hours and 50 minutes.

The terms *tidal day* and *lunar day* are used interchangeably. A *tidal period* is the elapsed time between successive high or low tides. In the example described here, the tidal period is 12 hours and 25 minutes, but in some cases, described below, the tidal period may coincide with the tidal day.

In the scenario depicted above, the observer experiences two high (and two low) tides a day. Since the observer is on a line of latitude directly under the moon (the moon is directly over the equator, and the observer lives on the equator), the heights of successive high tides are equal. Such a tidal pattern is called a *semidiurnal tide* (Figure 8.20a). However, many locations on earth

FIGURE 8.19 The tidal day is 24 hours and 50 minutes long. (From Anikouchine, W. A., and Sternberg, R. W. 1973. *The world ocean.* Englewood Cliffs, N.J.: Prentice-Hall.)

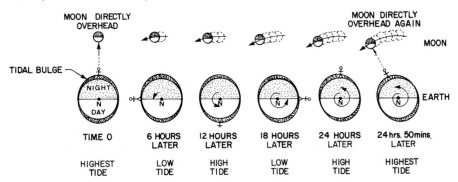

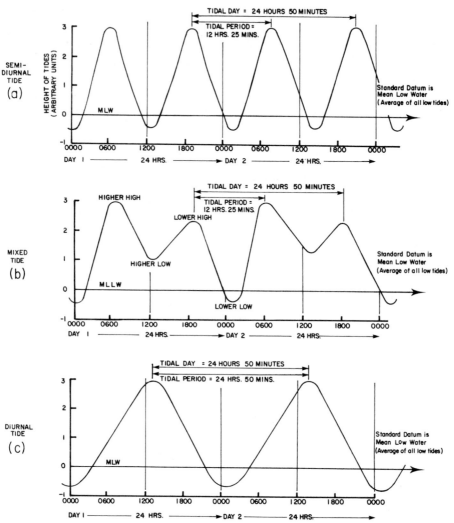

FIGURE 8.20. Semidiurnal, mixed, and diurnal tide plots. (From Weiss, H. M., and Dorsey, M. W. 1979. *Investigating the marine environment: a sourcebook.* Groton, Conn.: Project Oceanology.)

experience two daily high tides of unequal heights or even only one high tide per day. The explanation for this lies in the varying position of the moon with respect to the equator.

THE EFFECT OF THE MOON'S DECLINATION ON THE EARTH'S TIDES. A major factor affecting the height of the tide at any point on earth is the location of that point relative to the position of the moon. All other things being equal, points lying directly under the

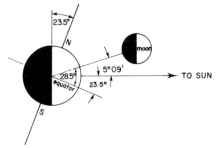

FIGURE 8.21. The declination of the moon reaches a maximum of approximately 28.5 degrees with respect to the earth's equator. (From Anikouchine, W. A., and Sternberg, R. W. 1973. *The world ocean.* Englewood Cliffs, N.J.: Prentice-Hall.)

moon during the course of a day will experience the highest tides on earth. The location of these points varies from day to day and from year to year because of the rather complex nature of the orbit of the moon around the earth. To understand this orbit, please refer to Figure 8.21. Note that the moon orbits the earth at an angle of up to 28.5 degrees with the equator. There are two reasons for this.

First, the earth is inclined on its axis at an angle of 23.5 degrees from the elliptic plane (*elliptic*) representing the orbital path of the earth around the sun. As a consequence the earth experiences seasons (Figure 8.22). Second, the orbit of the moon around the earth is itself inclined 5 degrees from the elliptic (Figure 8.21). The plane of this orbit is not stationary but instead rotates, or *precesses*, with a period of 18.6 years, always maintaining its 5-degree angle with the elliptic.

FIGURE 8.22. The earth experiences seasons because its axis of rotation is not precisely perpendicular to the plane of its orbit around the sun. (From Thurman, H. V. 1981. *Introductory oceanography.* 3rd ed. Columbus, Ohio: Merrill.)

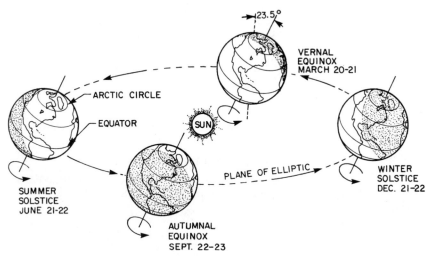

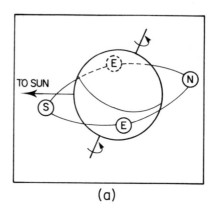

(a)

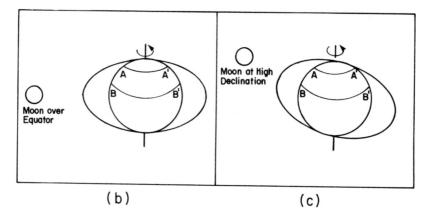

(b)   (c)

FIGURE 8.23. The declination of the moon affects the earth's tides. (a) During the lunar month, the moon is directly over the equator on two separate days (position E); at maximum north declination on one day (position N); and at maximum south declination on one day (position S). During the rest of the month it is in between a position over the equator and maximum declination. (b) The tidal bulge with the moon directly over the equator. (c) The tidal bulge with the moon at maximum declination. (From Weiss, H. M., and Dorsey, M. W. 1979. *Investigating the marine environment: a sourcebook*. Groton, Conn.: Project Oceanology.)

As a result of these factors, the angle of the moon above or below the equator (the moon's *declination*) is a maximum of 28.5 degrees every 18.6 years and a minimum of 18.5 degrees every 9.3 years. Remember, however, that the moon orbits the earth with a period of 30 days. Consequently, its path will carry it directly over the equator (and also to its maximum monthly declination) twice a month (Figure 8.23a). Thus, the highest tides on earth, ideally, occur along the plane of the moon's orbit. Any point on that plane will experience two equally high daily tides (semidiurnal tides) twice a month, when the moon is directly over the equator (Figure 8.23b).

During those times of the month when the moon is not directly over the equator, some peculiar tides can occur on the surface of the earth due to the influence of the moon alone. To simplify our understanding of the response of different locations to lunar tidal forces, let's tilt the earth-moon system so that the earth spins in an upright position and, for example, the moon's declination is 28.5 degrees north of the equator (Figure 8.23c).

Now imagine you are living at Point A and that you are directly under the moon at 0600 on Day 1. Twelve hours later, at 1800 on Day 1, you will have rotated to a position on the other side of the moon and will experience another high tide. But if you look closely at Figure 8.23c, you will note that the bulge experienced at time 1800 is less than that at 0600; your later high tide will be lower than the earlier one. At times 1200 and 2400 you will be midway between the high-tide points and will experience low tides, but the height of the two low tides will also be different. At 0600 on Day 2 you will again experience a high tide. The tidal pattern exemplified here is a *mixed tide*, in which there are two highs and two lows a day, as in a semidiurnal tide, but the heights are different. A mixed tidal curve is depicted in Figure 8.20b.

An even more interesting situation would occur if you were living at Point B. As at Point A you would experience a high tide at 0600 on Day 1; however, at time 1800 you would not be under even a slight bulge. Not until 0600 on Day 2 would you experience a high tide again. Only one high tide a day would occur at your location, and the tide would reach its lowest point 12 hours after a high tide. Such a tide is a *diurnal tide*, whose curve is depicted in Figure 8.20c.

The Distance of the Moon from the Earth

Before turning our attention to the influence of the sun on the tides, there is one more independent effect of the moon to consider: the distance of the moon from the earth. As the moon revolves around the earth it describes an elliptical and slightly offset orbit, rather than one which is completely spherical. At its closest approach to the earth during the month (*perigee*), the moon is about 375,000 kilometers away. At its furthest point (*apogee*), the moon is nearly

406,000 kilometers from earth. The lunar tide-generating force is greater at perigee than at any other time during the month.

Interacting Effects of the Sun and the Moon: The Effects of the Sun on Tides

The effects of the sun on the tides are of the same nature but lesser magnitude than those of the moon. While the gravitational attraction between two bodies is inversely proportional to the square of the distance between them, because of centrifugal force the total tide-generating force exerted by a celestial body on the earth is inversely proportional to the *cube* of the distance between them. Even though the sun weighs 27 million times as much as the moon, it is 390 times farther away. Since the cube of 390 is slightly more than 59 million, considering the distance factor alone, the moon's tide-generating effect is over 59 million times that of the sun. Considering the mass effect alone, the tide-generating effect of the sun is about 27 million times that of the moon. Dividing the latter by the former (27/59) reveals that the sun's effect on the earth's tides is only 46 percent that of the moon's.

The independent influence of the sun on the earth's tides is also affected by the variable distance between the two bodies. Just as the moon describes an offset, elliptical orbit around the earth, the earth's annual journey around the sun does not follow a perfect symmetrical circle. On January 2 the earth achieves its closest proximity to the sun (148,500,000 kilometers), and the independent effect of the sun on tidal range is the greatest of any time of the year. The sun is described as at *perihelion* with the earth at that time. On July 2 the earth is further from the sun than at any other time of the year (152,200,000 kilometers), and the sun's effects on the tides are least. This relative position of the sun is called *aphelion*.

Interactions of the Sun and Moon

Twice a month the sun, moon, and earth are in line (Figure 8.24). On one of these occasions the moon is between the earth and the sun (in *conjunction*), and a new moon is seen. Alignment of the three bodies also occurs when the moon is on the opposite side of the earth from the sun (in *opposition*), and the moon is full. With the moon in conjunction or opposition the tide-producing forces of the sun and moon are additive, and the *range* of the tides (the verti-

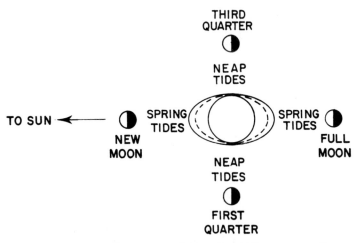

FIGURE 8.24. Interacting effects of the sun and the moon on the earth's tides. (From Weiss, H. M., and Dorsey, M. W. 1979. *Investigating the marine environment: a sourcebook.* Groton, Conn.: Project Oceanology.)

cal distance between high and low tide) is the highest of any time during the month. This means that the high tides ascend higher and the low tides recede lower than at any other time of the month. These tides are called *spring tides,* though they have nothing to do with the season of the year.

Neap tides also occur twice a month and mark the occasions when the tidal range is the lowest of the 30-day period. They occur when the sun and moon form a right angle with the earth (Figure 8.24). The moon is then said to be in *quadrature* (we see a *quarter moon*), and the tide-generating forces of the sun and moon work against each other.

The Equilibrium Theory of the Tides

Our discussion thus far has involved some important simplifying assumptions. These were originally delineated by Sir Isaac Newton, who was the first to propose a mathematical explanation for the tides, three centuries ago. The assumptions are that the ocean is static, that it completely covers the earth, and that its response to the tide-generating forces of the sun and moon is unimpeded by the existence of land masses. While we know that these assumptions are false, by making them Newton was able to explain tidal behavior surprisingly well on the basis of gravitational and centrifugal forces alone. He summarized this explanation in his *equilibrium theory of the tides.* This is essentially the theory described in the discussion to this point.

The Dynamic Theory of the Tides

While the equilibrium theory provides a reasonably adequate explanation for tidal behavior in many parts of the world, and even allows some roughly accurate predictions in those locations, it does not work at all in other places. This is because the movement of the oceans' waters in response to the tide-generating forces of the sun and moon is significantly modified by a number of *dynamic* factors on the earth's surface. These include the influences of the winds, the Coriolis effect, ocean currents, friction, and land masses on the movement of the waters. These influences are complicated and have taxed the imaginations and abilities of mathematicians and physicists who have been attempting to accurately predict tides since Newton's time. Their incorporation into tidal theory constitutes what is called the *dynamic theory of the tides,* and we shall discuss it briefly here.

FIGURE 8.25. An idealized tidal standing wave in a rectangular ocean basin with the wavelength equal to the basin width. The wave period is twelve hours. (From Fox, W. T. 1983. *At the sea's edge.* Englewood Cliffs, N.J.: Prentice-Hall.)

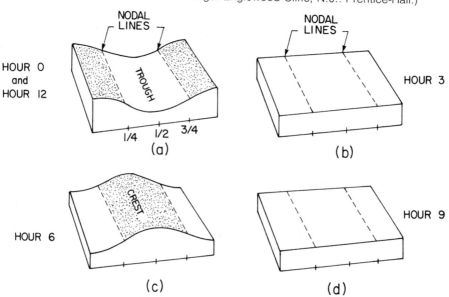

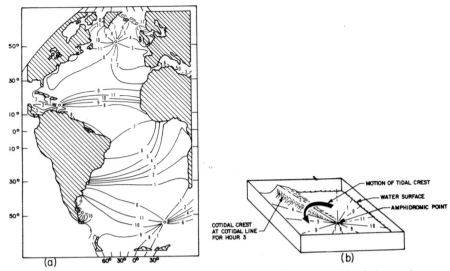

FIGURE 8.26. (a) Cotidal lines showing locations of tidal crest at hourly intervals over a semidiurnal tidal period. Amphidromic points are locations of no tidal fluctuation (zero tidal range) and occur where cotidal lines merge. (b) Water-surface displacement and motion of tidal crest around an amphidromic point. (From Anikouchine, W. A., and Sternberg, R. W. 1973. *The world ocean.* Englewood Cliffs, N.J.: Prentice-Hall.)

STANDING WAVES AND AMPHIDROMIC TIDES. Perhaps the most important dynamic factor affecting the tides is the presence of continental land masses. Continents essentially divide the ocean into separate basins and prevent the water in these basins from moving completely around the earth as an unobstructed wave. When a tide-generated wave reaches the western boundary of an ocean basin, its leading edge is reflected back by the continent even as the remainder of the wave is still moving toward the land mass. This creates a *standing wave* (Figure 8.25), with nodal and antinodal points and a natural period of oscillation (see also Figure 8.7). The period of oscillation and the number and location of nodal and antinodal points is a function of the length and depth of the ocean basin. You will recall that there is no vertical displacement of the sea surface at nodal points. Consequently, if you happen to live in a location (perhaps an island) corresponding to a node, you will not experience any tides (i.e., tidal range is zero). These locations are also known as *amphidromic points* (Figure 8.26).

Strong winds or seismic disturbances may set up standing waves in ocean basins which are independent of the forced wave produced by the tide-generating force in that basin. These *free-*

standing waves, as they are called, have their own characteristic period and oscillation pattern which, in certain circumstances, may be close to those of the tide-produced wave. The waves then reinforce each other, producing *resonance tides* with an abnormally large range. Free-standing waves in tideless lakes are known as *seiches.* Resonance is partly responsible for the extremely large tidal ranges encountered in the Bay of Fundy. At their maximum, in the upper end of the Bay, the range approaches 17 meters.

The behavior of the tides is also influenced by the Coriolis effect, described in Chapter 6. The effect is particularly pronounced in broad ocean basins where the deflection of water due to the earth's rotation causes the crest of the "tidal wave" to move in a rotary fashion around the periphery of the basin. Interaction of the Coriolis effect and a standing wave produced by a tide may produce additional amphidromic points.

Tides in narrow basins are strongly influenced by the constricting effect of the basin. A *flooding* (incoming) or *ebbing* (outgoing) tide will cause a large volume of water to be squeezed through a narrow passage. This in turn increases the velocity of the ebbing or flooding *tidal current* because a constant volume of water, representing the volume difference between the high and low tide levels, is exchanged during a tidal period. The narrower the opening through which that volume must flow, the faster the water must move to accommodate the volume during the fixed tidal period. The principle involved is the *venturi effect.* The magnitude of tidal currents is thus a function of tidal range and the width of tidal passages. It is also affected by winds, nontidal currents (e.g., incoming rivers), and frictional contact with the bottom (partly a function of water depth) and with the boundaries of basins.

In some parts of the world a combination of shallow coastal waters and a large tidal amplitude produces spectacular tidal waves called *tidal bores.* When the trough of the long wavelength tidal wave slows down because of the shallow depth of water, the crest of the tidal wave catches up, producing a steep wavefront which may reach heights of several meters.

Anyone who spends time on the water should know how to determine the time and heights of tides and the direction and strength of tidal currents. Fortunately, this information is readily available for most coastal locations in the United States (and many areas in the rest of the world). Predictions of the times and heights of high and low tides are the province of publications generically referred to as *tide tables*. In the United States these are published annually by the National Ocean Survey, an agency within NOAA, in four separate volumes, covering four different areas of the world. Predictions of the direction and magnitude of tidal currents are contained in various tidal-current charts and tables. The National Ocean Survey publishes two volumes of tidal-current tables each year. One covers the Atlantic Coast of North America and the other the Pacific Coast of North America and Asia.

SUGGESTED READINGS

Waves

BASCOM, W. 1959. Ocean waves. *Scientific American* 201:89-97.

BASCOM, W. 1964. *Waves and beaches.* Garden City, N.Y.: Doubleday and Doubleday.

ISELIN, C. 1963. The loss of the *Thresher. Oceanus* 6:4-6.

LAND, T. 1975. Freak killer waves. *Sea Frontiers* 21:139-141.

MOONEY, M. J. 1975. Tragedy at Scotch Cap. *Sea Frontiers* 21:84-90.

PARARAS-CARAYANNIS, G. 1977. The International Tsunami Warning System. *Sea Frontiers* 23:20-27.

SMITH, F. G. W. 1971. The real sea. *Sea Frontiers* 17:298-311.

SMITH, F. G. W. 1970. The simple wave. *Sea Frontiers* 16:234-245.

STOWE, K. 1983. *Ocean Science.* 2nd ed. Chapter 8. New York: Wiley.

SVERDRUP, H. U., JOHNSON, M. W., AND FLEMING, R. H. 1942. *The oceans: their physics, chemistry, and general biology.* Chapter 14. Englewood Cliffs, N.J.: Prentice-Hall.

THURMAN, H. V. 1981. *Introductory oceanography.* 3rd ed. Chapter 11. Columbus, Ohio: Merrill.

Tides

ANIKOUCHINE, W. A., AND STERNBERG, R. W. 1973. *The world ocean: an introduction to oceanography.* Chapter 9. Englewood Cliffs, N.J.: Prentice-Hall.

GOLDREICH, P. 1972. Tides and the earth-moon system. *Scientific American* 226:42-57.

SVERDRUP, H. U., JOHNSON, M. W., AND FLEMING, R. H. 1942. *The oceans: their physics, chemistry, and general biology.* Chapter 14. Englewood Cliffs, N.J.: Prentice-Hall.

THURMAN, H. V. 1981. *Introductory oceanography.* 3rd ed. Chapter 12. Columbus, Ohio: Merrill.

WEISS, H. M., AND DORSEY, M. W. 1979. *Investigating the marine environment: a sourcebook.* Pp. 387-392. Groton, Conn.: Project Oceanology, Avery Point.

Chapter Nine

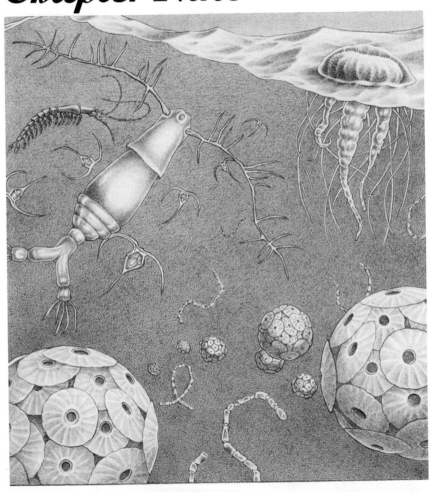

Life in the Sea I

NO LIVING CELLS PULSED ON THE BARREN LANDSCAPE OF THE early earth; no gliding forms graced its silent seas. As we marvel at the diversity, complexity, and adaptability of life on earth today, it is hard to believe that once our world was dead.

Despite such desolation, however, the ingredients of life were in place nearly 4 billion years ago. The new earth had cooled, and waters covered the globe. Sunlight streamed through a protective, enveloping atmosphere, energizing the world below. A potpourri of natural elements suffused earth, water, and sky. The potential for life lurked in the wings, waiting for an organizing force.

What forces were responsible for the life-giving organization of these ingredients? The answers still largely elude us, but we have gathered a few clues. They point strongly toward the ocean as the site for the origin of life on earth. There are two major pieces of evidence for this view.

First, consider the problem of ultraviolet (UV) radiation. Much of the energy of the sun is emitted in this portion of the electromagnetic spectrum. With a shorter wavelength and higher energy than visible light, direct UV radiation damages chromosomes and tissues of unprotected cells. Today, most UV radiation is absorbed in the ozone layer of the upper atmosphere; very little reaches the surface of the earth. However, there was no ozone or free oxygen in the early atmosphere. UV radiation streamed down, unobstructed, to the earth. It is unlikely that terrestrial life forms could have survived such a bombardment. Water, however, blocks UV radiation, indicating that the first living organisms must have evolved below the protective sea surface layer.

The oceans not only protected the first life forms; their constituents probably also produced them. Even the early seas were a soup of virtually all elements and inorganic chemicals existing on earth at the time. Since these materials are the essential components of all life forms, the ingredients of life were well dispersed and readily transported in the liquid medium of the oceans.

It is one thing to demonstrate that the early oceans harbored the materials and conditions appropriate for life to evolve and quite another to explain how living cells could be created from nonliving constituents. Again, we have some clues as to how this might have happened.

Three decades ago Stanley Miller assembled a unique apparatus in a laboratory at the University of Chicago. Essentially, the apparatus was a simplified model of the conditions believed to prevail in the early earth. A heated reservoir of water supplied vapor to a simulated atmosphere containing methane, ammonia, and hydrogen, all probable constituents of the ancient atmosphere. Miller hypothesized that, with a powerful source of energy, the inorganic compounds might be reassembled into complex *organic molecules*—carbon-containing compounds which are the building blocks of life itself. Miller hypothesized that lightning might have supplied such energy to the primitive world. Therefore, to simulate electrical storms, the scientist applied a periodic spark to his mixture. Within days, a number of organic molecules were created, including amino acids, the basic units of proteins. The experiment was no fluke; it has been repeated frequently in intervening years.

Remarkable as Miller's results were, they did not demonstrate the creation of life itself. A number of characteristics distinguish living matter from nonliving, organic molecules. Chief among these are the capacity for self-replication (enabled by DNA), active metabolism (including the exchange of materials with the environment), response and adaptation to a changing environment, growth, and a high degree of internal organization. Despite the concerted efforts of a large number of scientists, we still have no clear idea as to how the transition to life itself might have occurred. We are even further from replicating the process.

Energy for Life

The first living organisms were single cells. All living cells require a source of organic molecules to supply their internal energy needs. Today, a variety of organisms, collectively called *autotrophs*, or *primary producers*, can make these organic molecules themselves from inorganic constituents. Chief among these are *photosynthetic* life forms, including the green plants. By absorbing and using the energy of the sun, photosynthesizers make energy-rich organic carbohydrate molecules from carbon dioxide and water (Figure 9.1). Free oxygen gas (O_2) is a byproduct of this reaction. Similarly, *chemosynthetic* autotrophs, exemplified by several kinds of bacteria, obtain the energy to make organic molecules through chemical alterations of inorganic compounds. Unlike photosynthesizers, chemosynthetic organisms do not produce oxygen.

Many of the creatures on earth cannot make their own food directly but obtain their nutrition by eating autotrophs or otherwise ingesting or absorbing preexisting organic molecules. Such orga-

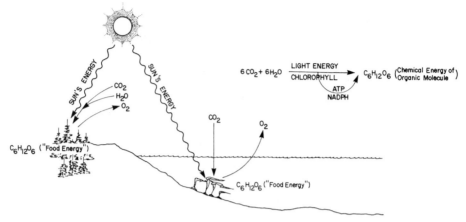

FIGURE 9.1. Photosynthesis.

nisms are known as *heterotrophs*. They include all animals and thus include ourselves.

Life today is dependent upon autotrophs. Without organisms that continually reassemble organic molecules from inorganic components, living forms would soon exhaust their food supplies. Life would grind to a halt.

Ironically, the first living creatures were probably heterotrophs, since the specialized metabolic mechanisms which enabled organisms to capture and use the energy of the sun were undoubtedly a long time in evolving. The evidence suggests that the earliest life forms were simple units of organized protoplasm, capable of self-replication, and that they absorbed organic molecules directly from the environment. The source of these molecules? In all likelihood it was the amino acids and other organic compounds created by the early earth conditions duplicated in the Miller experiment.

Early Evolution

Let us reconstruct the probable scenario for the first living cells on earth. Immersed in the seas (probably in shallow waters), they encountered a difficult dilemma. They were surrounded by a readily available food supply which they quickly exploited. As their numbers grew, their resources shrank. The demand for new organic molecules soon exceeded the supply from natural processes. The first living creatures were rapidly "eating themselves out of house and home." Extinction loomed menacingly ahead.

Fortuitously, inexplicably, an astonishing event occurred. Among the myriad, frequently debilitating genetic mutations continuously undergone by living cells then, as now, some proved to be beneficial and were retained by populations through natural selection. Among these was a sudden (in the geological sense of time) capacity for autotrophic metabolism, probably in the form of photosynthesis. At once new horizons were opened to life on earth. Organic molecules could be created as fast as they were consumed. Within a short fraction of the planet's history, life exploded across the earth.

The first photosynthetic organisms were probably single-celled ancestors of the *blue-green algae,* or *cyanobacteria.* (They are closely enough related to bacteria to be included within that group by many microbiologists). They were as susceptible to harmful ultraviolet rays as any of earth's creatures and were initially confined to the seas. However, as they evolved and multiplied, they profoundly altered the very environment which threatened them. Recall that free oxygen gas is a byproduct of photosynthesis. Not used directly in the reactions, it is released as a waste to the environment. The first autotrophs produced large quantities of oxygen which gradually accumulated in the atmosphere. There, much of it reacted with UV radiation to produce ozone. With the creation of an ozone shield, life was free to come ashore.

The accumulation of oxygen in the atmosphere had another benefit. Though this waste product may have been initially toxic to living cells, they eventually evolved the near-universal capacity to exploit the gas to their own advantage. Today, nearly all living organisms require molecular oxygen, either in the air or dissolved in the aquatic environment. The oxygen is needed to completely oxidize organic matter, thereby efficiently converting it to cellular energy, during the process of *cellular respiration.* (*Oxidation* is the process of removing electrons from a substance, resulting in the release of energy.)

LIFE IN THE MARINE ENVIRONMENT

The Marine Environment as a Habitat

Life today occupies every conceivable environment on earth, from the searing heat of the desert to the frozen wastes of high latitudes; from the crushing pressures of the ocean depths to the rarefied reaches of the upper atmosphere. Of all environments, the seas are by far the most benign.

Compare life in watery surroundings to a terrestrial existence. The danger of drying out is remote, obviating the need for energetically expensive protective structures for adult and juvenile life forms. The buoyant force of water provides physical support. Required chemical compounds are dissolved, well distributed, and readily available to aquatic organisms. The disposal of potentially toxic metabolic wastes is facilitated by the diluting and dispersal properties of a liquid medium. The ocean is a comparatively stable environment, with limited fluctuations in temperature and chemical characteristics. The broad distribution of populations is aided by currents and other water movements.

Limiting Factors and Adaptations

However benign the marine environment may be, life in the oceans is not easy. Consider the requirements of an aquatic organism. Temperature, salinity, and pH must lie within the range to which the organism is adapted. Essential nutrients and gases must be available in appropriate quantities. For photosynthetic organisms, there must be enough light to allow the necessary reactions to proceed. Seldom are all requirements available in exactly the right amounts and proportions. Almost always, the availability of one or more factors exceeds or falls short of optimal levels for a specific organism. Consequently, the organism does not achieve its maximum growth rate or reproductive potential. The responsible inhibiting factor is called a *limiting factor.*

Of the host of potentially limiting factors in the marine environment, there are several general categories which are most likely to affect resident organisms. These include temperature, salinity, pressure, light, living space, dissolved gases, nutrients or food, and factors affecting flotation. Two of these factors, temperature and salinity, merit special discussion here. The other factors will be considered when we examine specific marine environments and the remarkable forms and adaptations of their inhabitants.

TEMPERATURE. Temperature affects marine organisms both directly and indirectly. In general, the rate of cellular activities, including photosynthesis, respiration, growth, and development, increases with temperature. However, all organisms have *optimal* temperatures at which they operate most efficiently. Temperatures above or below optimal levels cause stress. High or low extremes of temperature (*upper* and *lower lethal limits*) cause death, while the temperature range within which an organism lives, but does not necessarily thrive, is called the *zone of tolerance*. The general response of living organisms, including marine species, to temperature is depicted in Figure 9.2.

The range of temperatures to which an organism is adapted varies with the species. Although the oceans are far more stable thermally than terrestrial environments, they still may undergo considerable variations in temperature. In general, coastal waters show greater fluctuations than offshore waters, deep waters have more constant temperatures than surface waters, and midlatitude marine environments experience greater annual temperature ranges than polar, subpolar, or tropical environments. Organisms inhabiting thermally stable waters tend to be well adapted to relatively constant temperatures and do not tolerate extremes well. Such species are called *stenothermal* and are exemplified by deep-sea fish. In contrast, organisms adapted to thermally fluctuating environments are termed *eurythermal*. Examples include most animals living in shallow, coastal environments.

For many animals, internal temperature mirrors that of the outside environment. Such animals, exemplified by reptiles and fish, are called *poikilotherms* (i.e., they are "cold-blooded"). Poikilothermy accounts for the phenomenon of snakes and lizards basking in the sun after a cold night; in a sense they are recharging their in-

FIGURE 9.2. Generalized response of an organism to temperature.

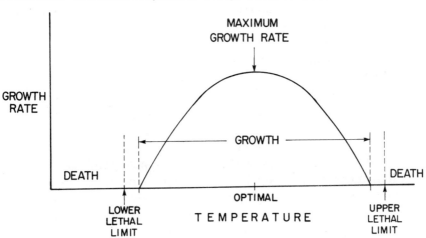

ternal metabolic "batteries." Other animals expend varying amounts of energy to maintain a relatively constant internal environment. This process, called *homeostasis*, extends to the maintenance of essentially unvarying inner temperature in mammals. Consequently, mammals are regarded as *homeotherms* (i.e., they are "warm-blooded").

Most animals in the sea are poikilotherms. The only true homeotherms in the ocean environment are the marine mammals (exemplified by whales, dolphins, and seals) and the sea birds.

Temperature also has indirect effects on marine organisms. The density and *viscosity* (flow-resisting properties) of water, and its capacity to hold vital gases like carbon dioxide and oxygen, are all increased by reductions in temperature. These relationships are elucidated in Chapters 5 and 7. In general terms, density affects the stability of the water column and the stratification or vertical movement of water masses. This in turn affects the availability of nutrients and the distribution of substances and organisms in the ocean. Viscosity affects the flotation and sinking rate of organisms and food particles.

SALINITY. Like temperature, the salinity of the marine environment is most variable in shallow, coastal waters and in the upper surface layers of the ocean. Where the land meets the sea it is not unusual for salinity to fluctuate between that of fresh water (0 parts per thousand) and fully oceanic (close to 35 parts per thousand), even achieving *hypersaline* (above-average salinity) conditions. Organisms successfully inhabiting such environments have adapted to these extremes and are referred to as *euryhaline*. Organisms adapted to the relatively constant salinities of deep, offshore environments are called *stenohaline*.

Before considering the ways in which marine organisms adjust to varying salinity, let's examine some basic principles of salt and water balance in living cells. Cell membranes are *selectively permeable* in that they may freely allow the passage of relatively small molecules, such as water, yet restrict or regulate the influx and efflux of ions and larger molecules, such as dissolved salts.

Unless restricted by a barrier, substances tend to move, by random molecular movement, down a *concentration gradient* (from a region of higher concentration to a region of lower concentration). Net movement ceases when the molecules achieve *equilibrium concentration* (i.e., are uniformly distributed). For substances other

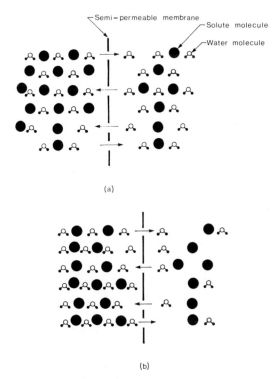

(a)

(b)

FIGURE 9.3. Osmosis.

than water, the term for this net movement is *diffusion* (see Chapter 5). *Osmosis* refers to the special case of water movement down a concentration gradient and through a semipermeable membrane (Figure 9.3).

Osmosis has important consequences for marine life. Consider what happens when aquatic organisms are located in a watery environment in which the concentration of dissolved solids (i.e., salinity) differs from that inside the cells. Water will osmotically flow through the cell membranes in response to its concentration gradient; however, the movement of solutes is restricted by the membranes. In consequence, the achievement of osmotic equilibrium is almost entirely up to water. This can mean an intolerably large loss or gain of water by the animal or plant.

Few marine creatures inhabit waters with a salinity identical to the organism's internal solute concentration. How then do they adjust to differing or fluctuating salinities? There are a variety of mechanisms. Some simply avoid problems by closing themselves up in impermeable shells or cell walls when an imbalance with the out-

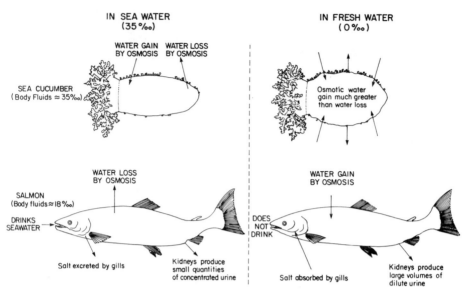

FIGURE 9.4. A sea cucumber and a salmon cope differently with osmotic imbalance with the external environment. (After Sumich, J. L. 1984. *An introduction to the biology of marine life*. 3rd ed. Dubuque, Iowa: Brown.)

side salinity creates a stress. Examples include numerous species of bivalve mollusks. Others, called *osmotic conformers*, allow water to freely flow in and out of their cells. Eventually, internal fluids become *isotonic* (having an equal solute concentration) with external sea-water salinity. (In comparing environments on the basis of solute concentrations, two additional terms are used: The *hypertonic* environment contains the higher solute concentration and the *hypotonic* environment the lower solute concentration.) Examples of osmotic conformers include some marine worms. To a large extent they are able to tolerate the significant variations in internal ion concentration which they often experience.

Less tolerant osmotic conformers include many sea cucumbers. Their internal fluids are virtually isotonic with offshore seawater. Since this is their normal environment, they do not experience osmotic problems. However, if a sea cucumber is placed in low-salinity water, the consequences can be disastrous. To achieve osmotic conformity, water floods into the animal, eventually swelling it to the near-bursting point (Figure 9.4). Severe tissue damage or death results.

Most marine organisms exhibit some degree of homeostatic control over the composition and concentration of their internal body fluids. Known as *osmoregulators*, they expend energy in regulating internal concentrations of salts and water. As a result they are able to tolerate considerable changes in external salinity. Salmon provide a fitting example. Since they are *anadromous* (reproducing in fresh water but spending most of their lives at sea), they encounter and tolerate an extraordinary range of salinities during their lifetimes. Their body fluids have a solute concentration about midway between that of fresh water and sea water. Without osmoregulatory mechanisms they would dehydrate in sea water and swell like sea cucumbers in fresh water. How do they avoid this? When swimming in a hypertonic environment, where they constantly lose water through osmosis, they restrict their waste output to small amounts of very salty urine, excrete salt from their gills, and drink lots of water to offset that lost by osmosis (Figure 9.4). When in a freshwater environment, they excrete large amounts of dilute urine, absorb salt through their gills, and forego drinking.

Biological Zones in the Oceans

Homogeneous as the ocean may seem to the casual observer, it is in reality a complex and changeable environment with a bewildering variety of habitats. While this extraordinary physical diversity defies easy classification, there have been numerous attempts to organize the oceans' living spaces into definable regions. One of the more widely accepted schemes is depicted in Figure 9.5. Note particularly the distinction between the *pelagic* (the water column of the open sea) and *benthic* (sea floor) environments.

The oceans and seas comprise over 90 percent of the earth's habitable environment, but they harbor far fewer species than does land. This is particularly true for the animal kingdom; terrestrial regions account for nearly 85 percent of all known species. Three quarters of these are insects; however, if we ignore this class, about two thirds of all animal species are marine.

Of 250,000 species in the oceans, two thirds of which are animals, 98 percent live on or in the bottom. They are referred to as the *benthos*. Pelagic (i.e., drifting or swimming) organisms, while less diverse, are better known and play an exceedingly important role in the seas' ecology.

Benthic environments will be discussed in detail in Chapters 10 (deep-sea environments) and 11 (shallow coastal environments) and deep-sea pelagic environments will be described in Chapter 10.

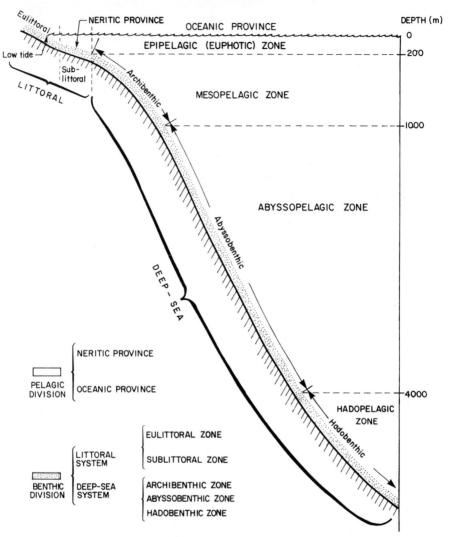

FIGURE 9.5. Major divisions of the ocean environment. (After Sverdrup, H. U., Johnson, M. W., and Fleming, R. H. 1942. *The oceans.* Englewood Cliffs, N.J.: Prentice-Hall.)

The remainder of this chapter will focus on the pelagic environment of the oceans' illuminated surface waters.

LIFE IN THE PELAGIC ZONE: THE OCEANS' SURFACE WATERS

The Surface-Water Environment

With this background, let's fully immerse ourselves in the marine environment. We'll first take a close look at the oceans' surface waters, focusing on their dominant life forms. We'll examine the conditions with which these organisms must cope, their diverse and remarkable adaptations, and their interrelationships. To a large degree, the upper layers of the seas drive all of the oceanic ecosystems, so it is fitting that our investigations begin there.

What are the limits of the oceans' surface waters? In essence they constitute the waters which are both well mixed and well illuminated. Refer to the diagram in Figure 9.5. The surface waters constitute the *epipelagic* or *euphotic* (sufficient illumination for photosynthesis) *zone.*

The epipelagic zone is a favorable environment for plankton and nekton. The availability of light ensures primary productivity. Mixing by waves and winds results in reasonably uniform dispersion of organisms, substances, and chemical and thermal properties. These conditions usually enable a steady production of food in the upper layers of the seas. Virtually all other organisms in the oceans ultimately depend on this food source, even though the surface waters constitute only about 2 percent of the ocean environment.

Our investigations will commence on a calm, sunny day in early spring in the surface waters of the North Atlantic Ocean, near the outer extremity of the continental shelf. As you ease yourself into the chilly waters, imagine that you are blessed with extraordinarily acute vision, enabling you to see microscopic and macroscopic life forms with equal clarity and detail. What kind of a world is revealed to you?

Plankton

Your first impression is of silent space saturated with fine, dustlike particles. In all directions this thick pall scatters and refracts the rays of the sun, making it difficult to determine the direction of the light source. On closer inspection, the particles turn out to be a large population of minute living organisms, densely suspended in the gently undulating swells. These are *plankton,* aquatic life forms which drift or float at the mercy of the waves and currents. Plankton are unique to the aquatic environment. Finely scattered through the water column, they need no surface on which to cling, no wings or fins to stay afloat. They are born, grow, reproduce, and die without ever coming in contact with solid ground.

Why should there be plankton when nothing like them exists in the fluid atmosphere of the terrestrial environment? Primarily, they exist because they can float. Living protoplasm is only slightly denser than sea water. Planktonic organisms, because of their high

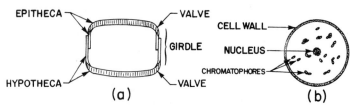

FIGURE 9.6. General structure of a centric diatom. (a) Cell wall in girdle-view. (b) Valve view of cell. (From Sverdrup, H. U., Johnson, M. W., and Fleming, R. H. 1942. *The oceans.* Englewood Cliffs, N.J.: Prentice-Hall.)

surface-area-to-volume ratio, have little difficulty remaining suspended. Energy which might otherwise be directed toward maintaining position in the water is channeled into growth and reproduction.

Phytoplankton

Examine the plankton closely. Most of them are *phytoplankton,* microscopic photosynthetic algae found in virtually all water bodies, including fresh water. What kinds of organisms are represented by this group?

DIATOMS. Your observations reveal that one phytoplankton form is particularly abundant. Look at it carefully. It is a single-celled alga with a transparent, rigid cell wall constructed of silica—the material of which glass is made. This microscopic phytoplankter literally lives in a glass house. The cell wall (called a *frustule*) consists of two sections called *valves.* One valve fits inside the other much like the lower section of a container fits inside its lid. For this reason diatoms are said to resemble miniature pillboxes. The diatom's protoplasm and important cellular organelles (e.g., the nucleus and the pigment-containing photosynthetic units called the *chloroplasts*) are located inside the frustule (Figure 9.6).

The frustules of diatoms are primarily responsible for the extraordinary and beautiful variety of forms characteristic of this group (Figures 9.7 and 9.8), a variety quite evident in the numerous species surrounding you. Most of the offshore species represented here are radially symmetrical (*centric diatoms*), existing primarily as modifications of squares, circles, or triangles. Closer to shore, another major group, the *pennate diatoms*, would be well represented. These forms are bilaterally symmetrical and are usually rod-shaped

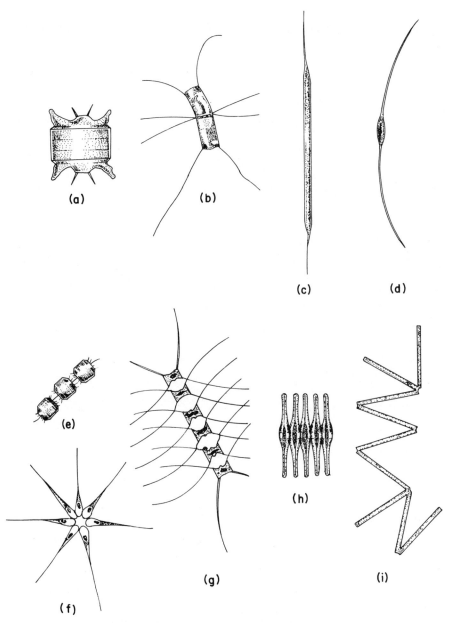

FIGURE 9.7. Representative diatoms. (a) *Biddulphia aurita*. (b) *Chaetoceros similis*. (c) *Rhizosolenia setigera*. (d) *Nitzschia longissima*. (e) *Thalassiosira nordenskioldii*. (f) *Asterionella japonica*. (g) *Chaetoceros didymus*. (h) *Fragilaria* sp. (i) *Thalassionema nitzschioides*. (Redrawn from Wood, R. D., and Lutes, J. 1968. *Guide to the phytoplankton of Narragansett Bay, Rhode Island.* Peacedale, RI: The Kingston Press. Copyright by Richard D. Wood.)

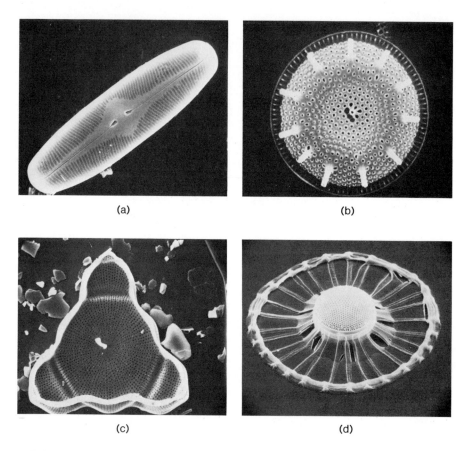

FIGURE 9.8. Photographs of frustules of representative diatoms. (a) *Navicula* sp., ×1089. (b) *Thalassiosira nordenskioldii,* ×2723. (c) *Lithodesmium* sp., ×1029. (d) *Planktoniella sol,* ×591. (Photographs by Paul Hargraves).

or longitudinal in their dimensions. While centric diatoms are truly planktonic, many pennate diatoms are benthic and capable of limited gliding motion along a surface.

Diatom frustules are resistant to decomposition and frequently accumulate on the sea floor below productive overlying surface waters. In some cases the accumulated skeletons form large deposits on the bottom, constituting most of the sedimentary material in the region. In fact, the mining of "diatomaceous earth" has become a profitable enterprise in some areas because the sediment's finely abrasive nature makes it an excellent polishing and cleaning agent. It is also used as a filter and deodorizer.

Growth and Reproduction of Diatoms. Unlike that of multi-cellular plants and animals, the "growth" of single-celled diatoms and other phytoplankton is not characterized by much expansion in individual size. Rather, it is manifested by increases in population which are achieved mainly through a process of simple *asexual* (without exchange of genetic material) cell division called *binary fission* (Figure 9.9). Under optimal conditions, diatoms divide at least once a day. At that rate a single parent cell would give rise to a million daughter cells in a three-week period.

Diatoms are confronted with a unique problem during cell division because their rigid frustule places constraints on simple binary fission. Perhaps to conserve the element silicon, diatoms retain the original valves but grow new valves inside them before the cell splits (Figure 9.9). Thus the original valve becomes the larger valve for each of the daughter cells. The result is that one new diatom is the same size as the parent, but the other is smaller. Upon maturity, the daughter cells reproduce in the same manner. Now only one fourth of the cells are the same size as the original diatom (Figure 9.9). The repetition of the process results in a progressive diminution in average diatom size.

How is a diatom version of *reductio ad absurdum* avoided? Diatoms smaller than a threshold size limit may undergo sexual reproduction leading to the formation of a *zygote* (the product of the

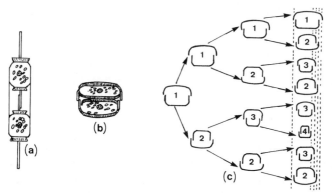

FIGURE 9.9. Cell division in diatoms. (a) and (b) Dividing cells. (c) Cell divisions in three successive generations result in marked diminution in average diatom size. (From Sverdrup, H. U., Johnson, M. W., and Fleming, R. H. 1942. *The oceans.* Englewood Cliffs, N.J.: Prentice-Hall.)

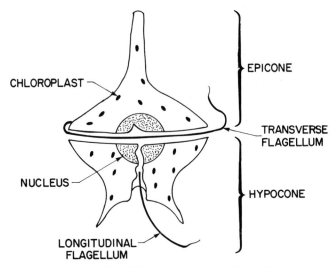

FIGURE 9.10. Idealized dinoflagellate showing major structures.

union of a male and female sex cell). The zygote lacks a frustule. The zygote then enlarges to become a naked *auxospore* which, under appropriate environmental conditions, may germinate, producing a new frustule and subsequently undergoing asexual cell division again.

As with other living organisms, *sexual reproduction* in diatoms confers variety in the population because the offspring contain the combined genetic characteristics of two parents. Variety, in turn, confers greater adaptibility of the population to a changeable environment.

OTHER PHYTOPLANKTON. While diatoms are the most abundant representatives in our population of floating algae, they are by no means the only phytoplankton in the sea. Let's examine some of the other forms in our area.

One interesting cell is about the same size as a diatom but lacks the characteristic frustule. As you watch it closely, you note that it is motile, whirling like a top before your eyes. Is this really a phytoplankter or a minute animal? The answer is supplied by examination of the cell's interior. Like all photosynthetic algae and unlike any animal, this individual contains intracellular chlorophyll. Its motility is due to a pair of whiplike appendages called *flagella* (Figure 9.10). This creature is a member of the taxonomic division Pyrrophyta, popularly called *dinoflagellates* because of the whirling motion imparted by the beating of its transverse flagellum (*dino* meaning "whirling"). The "animal/plant" confusion created by this group is exacerbated by the fact that some dinoflagellates even lack chlorophyll, instead directly absorbing organic compounds from the surrounding water.

Dinoflagellates are often as productive as diatoms and, under certain conditions, account for most of the primary productivity in a body of water. In their diversity of shapes and overall beauty they rival the diatoms (Figures 9.11 and 9.12). Some forms ("armored")

FIGURE 9.11. Representative dinoflagellates. (a) *Gymnodinium nelsoni.* (b) *Dinophysis acuminata.* (c) *Polykrikos* sp. (d) *Ceratium longipes.* (e) *Ceratium fusus.* (f) *Peridinium depressum.* (g) *Ceratium tripos.* (Redrawn from Wood, R. D., and Lutes, J. 1968. *Guide to the phytoplankton of Narragansett Bay, Rhode Island.* Peacedale, RI: The Kingston Press. Copyright by Richard D. Wood.)

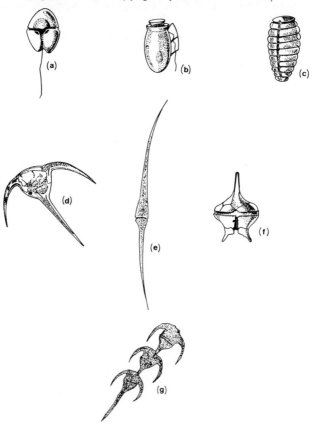

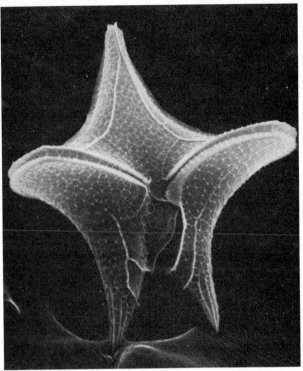

FIGURE 9.12. A representative dinoflagellate. *Peridinium* sp. (Photo by Paul W. Johnson and John McN. Sieburth, ×740).

are covered with overlapping plates made of cellulose. Other forms ("naked") lack these cellulose plates. In armored dinoflagellates the plates are frequently arranged into bizarre projections resembling wings, horns, or spines (Figures 9.11 and 9.12). These projections are probably adaptations to increase surface area to volume ratio to facilitate buoyancy or nutrient absorption. This is discussed in greater detail below.

Like the diatoms, dinoflagellates primarily reproduce through binary fission, but cell division is longitudinal. The cell regrows quickly, and no diminution in size occurs. Sexual reproduction apparently also occurs.

If your observations were taking place after dark, your movement through the water might well stir up an eery luminescent glow. This faint light would be attributable to *bioluminescence*, the production of light by living creatures. Bioluminescence is widespread in the oceans and is also practiced by some terrestrial organisms, most notably the fireflies. In the seas, chief practitioners include various species of dinoflagellates.

Dinoflagellates may be beautiful and ostentatious, but a few notorious representatives have given the group a bad press. About a dozen species produce extraordinarily toxic compounds, most of which act directly on the nervous systems of vertebrates (animals with backbones). One particularly bad agent, the species *Gonyaulax tamarensis* (Figure 9.13), has been responsible for the frequent closing of shellfish beds in New England. Each cell produces a minute quantity of a neurotoxin called *saxitoxin*. When a sufficiently large bloom (a million cells per liter of sea water or more) of these organisms occurs, under conditions still largely undefined, their pigments give the water a reddish hue. Consequently, such blooms are called "red tides." (Toxic red tides are also found in other parts of the world and are attributable to other species of algae. "Red tides" are also associated with large populations of some nontoxic marine organisms, so an indication of reddish water is not always a cause for alarm.)

FIGURE 9.13. The dinoflagellate *Gonyaulax tamarensis,* responsible for the "red tide" in New England. (Photograph by Paul Hargraves, ×2774)

When a red-tide bloom of toxic dinoflagellates occurs, shellfish (primarily mollusks) ingest large quantities of these microscopic algae in the process of indiscriminately filtering food particles from the water. The toxin accumulates in the tissues of these *filter-feeders* but does not harm the animals because, as *invertebrates* (without backbones), they lack complex nervous systems. However, the unwary vertebrate, including any human, which eats contaminated clams, mussels, or oysters is in for a rough time. Since saxitoxin is a neurotoxin, relatively small levels (saxitoxin is 50 times as deadly as its more famous cousins, curare and strychnine) cause muscular paralysis and death in vertebrates. Thus, food poisoning caused by saxitoxin is referred to as *Paralytic Shellfish Poisoning* (PSP). PSP has been responsible for several hundred deaths worldwide.

It has been long assumed that the diatoms and dinoflagellates are far and away the most productive and abundant phytoplankton in the seas. This assumption has primarily been based on the fact that traditional methods of sampling plankton have involved towing fine-meshed nets through measured volumes of water and then identifying and enumerating the captured organisms. The organisms thus captured are referred to as the *net plankton*.

Since the finest effective phytoplankton nets generally have mesh sizes of at least 60 microns (a micron is 0.001 millimeter or about 0.00004 inch), photosynthetic plankton smaller than a small

FIGURE 9.14. Coccolithophore. (Photograph by Paul Hargraves, ×7725)

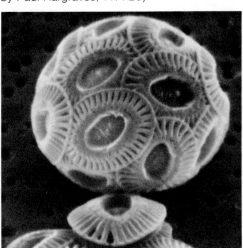

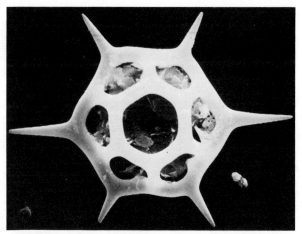

FIGURE 9.15. Silicoflagellate. (Photograph by Paul Hargraves, ×2145)

dinoflagellate or diatom pass through the pores. As a result, ocean-ographers have largely neglected a significant fraction of the oceans' primary producers. This fraction, considered to include forms smaller than 60 microns in diameter, is generally referred to as the *nanoplankton*. Though their importance has been suspected for decades, it has proved exceedingly difficult to study them because they are so small and hard to preserve.

Among the nanoplankton are two important groups which possess hard skeletons. The *coccolithophores* have a number of intricate and delicately sculpted external calcareous plates comprising their cell walls (Figure 9.14). They are widespread in tropical and temperate waters and are particularly abundant in the Sargasso Sea, where recent evidence suggests that they may be responsible for most of the primary production. The other group, the *silicoflagellates*, is characterized by elaborate internal skeletons made of silica (Figure 9.15). They are an obscure group, and it is not believed that they contribute significantly to oceanic primary productivity.

Recently, new techniques have revealed surprisingly numerous microscopic algae which are even smaller than coccolithophores and silicoflagellates. These bacteria-sized (0.2 to 2.0 microns), photosynthetic plankton are considered to be a new size class, the *picoplankton*. Identification and quantification of these tiny ocean denizens have proven difficult, but it appears that many are small species of blue-green algae (cyanobacteria) or single-celled green algae (Figure 9.16). Preliminary research indicates that they proba-

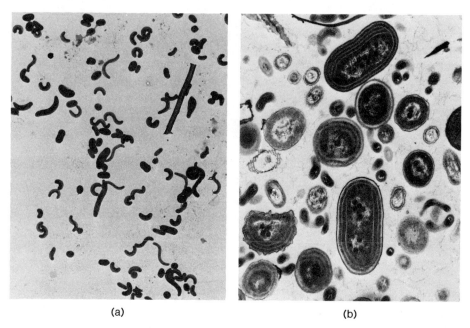

FIGURE 9.16. Picoplankton. (a) Morphological diversity of the bacteria in the picoplankton of the Gulf Stream, ×6400. (b) Picoplankton from the Gulf Stream containing heterotrophic bacteria (small cells) and chroococcoid cyanobacteria (large cells with thylakoids), ×17,920. (Photographs by Paul W. Johnson and John McN. Sieburth.)

bly contribute significantly to primary productivity in the seas—even in offshore waters, where total productivity has always been assumed to be low.

FLOTATION OF PHYTOPLANKTON. A phytoplankton cell is well suited to its pelagic existence. For one thing, its small size maximizes its capacity to exchange materials with the environment (a function of its outer surface area) relative to its internal needs (a function of its volume). It can be readily demonstrated that, given the same shape, the smaller an object, the larger its *surface-area-to-volume ratio* (Figure 9.17).

Small size and a large surface-area-to-volume ratio also facilitate flotation. Watch a few diatoms closely, still imagining that you have extraordinarily acute vision. On a calm day they waft slowly and gracefully through the water column. Even the slightest amount of turbulence is usually sufficient to toss them back up toward the water's surface. It does not require a practiced eye to note that phytoplankton experience a precarious condition between sinking and floating. This is well and good.

Floating (i.e., buoyancy or suspension) enables the photo-synthetic cell to remain within the illuminated zone of the water column where it can be productive. How do passively drifting phy-toplankton stay in the epipelagic zone, even though they are slightly denser than water? Some forms resemble plates or flat discs (Figure 9.7), shapes which increase drag. Aggregation into colonies and the presence of elaborate appendages and spinous processes in other species (Figure 9.7) may also confer resistance to sinking by in-creasing surface-area-to-volume ratio. Several phytoplankters pro-duce and retain light fats and oils which help them to float. Some diatoms evidently regulate their internal ratio of heavy and light ions, which may be a way to control flotation.

The ability of phytoplankton to remain in the euphotic zone is also affected by physical factors, including the viscosity of the sea water. Increased viscosity results in greater resistance to sinking and is affected by temperature and salinity. Temperature has a more pronounced effect. At 0° C sea water is twice as viscous as at a tem-perature of 25° C. Ocean water is 6 to 9 percent more viscous than fresh water. Finally, the turbulence produced by waves in the sur-face waters helps to keep phytoplankton suspended in the upper layers of the seas.

Despite the necessity to float, some sinking is beneficial to phytoplankton, since it exposes the cell to new sources of nutrients. Furthermore, the slight turbulence at the cell surface caused by movement through the water increases the planktonic cell's absorp-

FIGURE 9.17. Demonstration of the effect of size on surface area to volume ratio. (a), (b), and (c) all have the same total volume, but (b) has twice as much surface area as (a), and (c) has four times as much surface area as (a) and twice as much as (b). Thus, the surface area to volume ratios are 1.5:1 for (a), 3:1 for (b), and 6:1 for (c). (From Curtis, H., and Barnes, N. S. 1981. *Invitation to biology.* 3rd ed. N.Y.: Worth, p. 62 [Figure 4-10]).

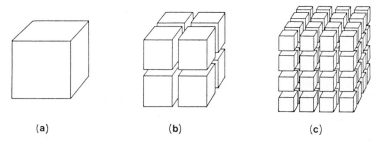

(a)   (b)   (c)

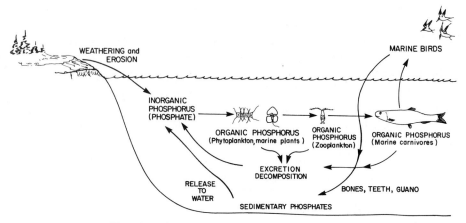

FIGURE 9.18. The phosphorus cycle in the marine environment.

tion rate of required gases and nutrients. One might therefore conclude that, for a phytoplankton cell, the best situation is not to float but to sink at an optimum rate (i.e., slowly enough to allow sufficient time in the euphotic zone for growth and reproduction).

PHYTOPLANKTON PRODUCTIVITY. As you glide through the top few meters of the water, you are astonished at the enormous quantities of algal cells around you. Clearly, productivity is at a peak now in the early spring. Never have you seen a concentration of living organisms as dense as the phytoplankton soup enveloping you. While the individual organisms would not be revealed to normal vision, the totality of this algal "bloom" is immediately obvious even to an observer out of the water: The algal pigments have turned the water brownish green.

What conditions account for such a vast population of phytoplankton? Given sufficient light, the most important factor limiting oceanic primary productivity is nutrient availability. Nutrients are the dissolved fertilizers of the ocean environment (see Chapter 5). They consist of the minerals, trace elements, and organic substances which photosynthetic organisms universally require in order to make important cellular chemicals and materials.

NUTRIENTS AND PRIMARY PRODUCTIVITY. The most abundant elements contained in the organic molecules of phytoplankton are hydrogen (H), oxygen (O), carbon (C), nitrogen (N), and phosphorus (P). The average approximate proportions of these are 230 H: 110 C: 75 O: 16 N: 1 P. Because of mixing and diffusion from the atmosphere, H, C, and O are usually available in ample supplies in sea water. N and P frequently are not. The comparative scarcity of these elements limits the total production of oceanic primary producers.

Phytoplankton absorb dissolved inorganic phosphorus primarily in the form of phosphate (PO_4^{3-}). In contrast to the situation for C, H, and O, the atmosphere does not contain significant amounts of P. Its availability to marine organisms is primarily dependent on runoff from land and the breakdown and recycling of P formerly contained in living cells (Figure 9.18).

Phytoplankton require about 16 times as much N as P. Since the demand for nitrogen frequently exceeds its supply in the oceans' surface waters, it is most often the factor which limits phytoplankton production in the euphotic zone. Ironically, the atmosphere is an enormous reservoir of nitrogen—over 78 percent of its constituents consists of N_2 gas. However, this form of nitrogen cannot be used directly by most primary producers. It must first be combined with hydrogen or oxygen into *fixed* N forms, including nitrate (NO_3^-), nitrite (NO_2^{2-}), and ammonium (NH_4^+) before it can be absorbed by algae and higher plants.

Certain *N-fixing* bacteria and blue-green algae (Figure 9.19) can accomplish this conversion efficiently enough to make N avail-

FIGURE 9.19. Cyanobacteria. The large clear structures are heterocysts, sites for nitrogen fixation. (Photograph by James Sears).

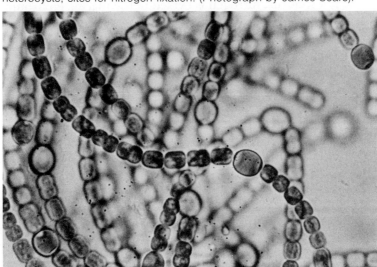

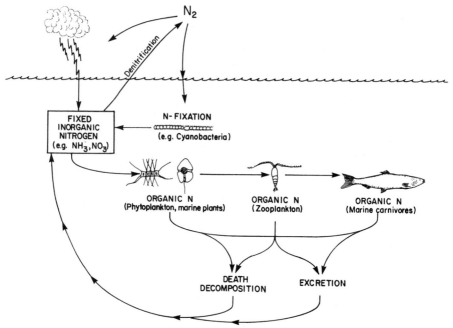

FIGURE 9.20. The nitrogen cycle in the marine environment.

able to other primary producers. Familiar terrestrial examples include the bacteria which inhabit the lumpy *nodules* on the roots of leguminous plants like peas and beans, and which reduce their need for nitrogen fertilizers. Some physical phenomena, including electrical discharges during thunderstorms, also accomplish N-fixation.

Like phosphorus, nitrogen is cycled in the marine environment (Figure 9.20).

Now reconsider our teeming North Atlantic surface waters. The dense population of phytoplankton indicates that nutrients are relatively abundant at the present time. This is largely attributable to the absence of a thermocline in the early spring (see Chapter 7). Without a thermocline there is no physical barrier to mixing of the nutrient-rich deeper waters with the overlying surface waters. In contrast, in the summer a layer of warm surface water traps colder water below the seasonal thermocline which develops in temperate latitudes.

If we were swimming in shallower waters closer to shore, phytoplankton productivity would likely be even greater, in part because shallow waters are even more readily mixed. Furthermore, coastal waters are a natural sink for the materials, including nutrients, which wash from surrounding land masses. In fact, nutrient runoff from human sewage outfalls and agricultural activities frequently causes large nuisance blooms of algae in some coastal environments. Large populations of phytoplankton in productive coastal waters are usually characterized by fewer species (i.e., lower *diversity*) than populations in offshore waters.

Zooplankton

Reluctantly, we tear our gaze away from the diverse and beautiful phytoplankton and focus on the larger, mostly microscopic, creatures which have periodically floated or darted into view. Most of these nonphotosynthetic organisms are *zooplankton*—the floating, drifting animals of the sea.

The zooplankton constitute an extraordinarily diverse group representing a variety of sizes, shapes, and lifestyles. Many are larval forms of larger, swimming adults. As you gaze at the bewildering array of tiny animals around you, your fascination mingles with confusion—how can anyone make sense of such a taxonomic smorgasbord?

THE COPEPODS. As you examine the zooplankton a little more closely, you note that one form is particularly abundant. Try to describe it. It somewhat resembles a miniature shrimp. Like a shrimp, it possesses an external shell and a number of jointed, clawlike appendages. Quick reference to a basic biology text would tell you that these are the primary characteristics of the arthropod phylum. (Please refer to Appendix II for taxonomic classification of principal marine organisms.) A little more research would reveal that, like shrimp, your new find is a member of the crustacean class. How do they differ from shrimp? For one thing they are much smaller, averaging between 1 and 5 millimeters in length—barely visible to the naked eye. They also have large, sweeping, oarlike antennae on their anterior ("head") end. These appendages are the most distinct features of this marine group—the *copepods* (subclass Copepoda).

As your observations suggest, copepods are enormously abundant and dominate the zooplankton through all of the earth's seas. The copepods are represented by numerous species, but one group, the *calanoid copepods*, is disproportionately abundant. In fact, one species of calanoid copepod, *Calanus finmarchicus* (Figure 9.21), has been called "the most important animal in the world" because of its huge populations in some regions.

Why are copepods important, other than by virtue of sheer numbers? Observe a few carefully. At first, they appear to be rela-

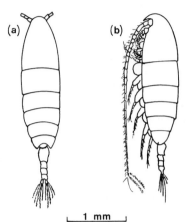

FIGURE 9.21. The copepod *Calanus finmarchicus*. (Redrawn from Newell, G. E., and Newell, R. C. 1963. *Marine plankton*. London: Hutchinson.)

tively motionless, minute lumps of suspended animation in the water column. A closer look, however, shows that these zooplankton are busy, rapidly moving the appendages on their ventral surfaces. In fact, they are feeding, by creating minute currents through the movement of their feeding appendages, from which tiny hairlike bristles (*setae*) project (Figure 9.22). As water is drawn toward the animal's mouth, food particles are removed and ingested.

Since these food particles are largely phytoplankton cells, the copepods are the principal grazers of the ocean environment. Thus, like terrestrial cattle, they are *herbivores* (i.e., they consume primary producers as their source of nutrition). Since many larger oceanic animals depend on zooplankton for their food, copepods

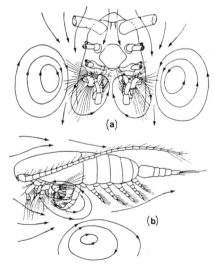

FIGURE 9.22. Depiction of filter feeding in the copepod *Calanus finmarchicus*. Arrows indicate direction of water movement and vortices. (a) Ventral view with distal ends of antennae and mandibles not shown. (b) Lateral view. (From Sverdrup, H. U., Johnson, M. W., and Fleming, R. H. 1942. *The oceans*. Englewood Cliffs, N.J.: Prentice-Hall.)

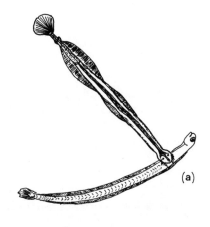

FIGURE 9.24. Arrow-worm swallowing a fish larva. (Adapted from Hardy, A. 1965. *The open sea: its natural history*. Boston: Houghton Mifflin.)

(b)

tinuous stream of water through openings at one end of its gelatinous "house." As the water flows out of holes at the other end, food particles are trapped by fine screens across the openings. Periodically, the resident animal sucks the food off a screen and ingests it. Understandably, the filtering device frequently clogs. In a matter of minutes, *Oikopleura* simply abandons its old house and builds another.

Among the smallest zooplankton are certain protozoans. Exceedingly abundant, surprisingly little is known about the ecological

FIGURE 9.25. *Oikopleura*. Arrows indicate direction of water flow through animal's "house." (From Hardy, A. 1965. *The open sea: its natural history*. Boston: Houghton Mifflin.)

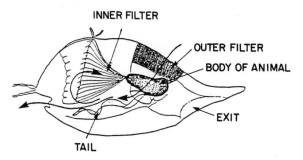

INNER FILTER

OUTER FILTER

BODY OF ANIMAL

EXIT

TAIL

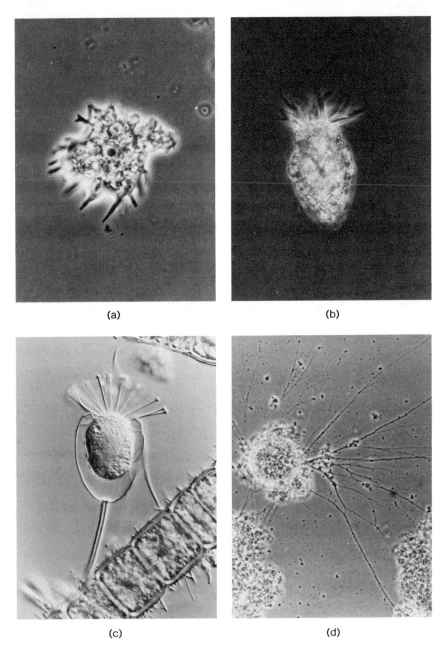

(a)

(b)

(c)

(d)

FIGURE 9.26. Marine protozoans. (a) A marine amoeba, *Mayorella* sp., ×589. (b) A tintinnid, ×512. (c) A ciliate suctorian, *Paracineta* sp., attached to a green alga, *Spongomorpha*, ×422. (d) A benthic foram, *Allogromia* sp., ×282. (Photographs by Paul W. Johnson and John McN. Sieburth.)

role of protozoans in marine environments. Oceanic protozoans include various species of amoebae, ciliates, and flagellates (Figure 9.26).

Oceanic amoebae range in size from 1 or 2 microns to over 5 millimeters in diameter. Many possess hard outer shells. Two groups of shelled amoebae are predominant in the oceans: the orders Foraminiferida (i.e., "forams") and Radiolaria ("radiolarians"). They feed by extruding their protoplasm through tiny perforations in the shell and surrounding and capturing tiny particles. When these organisms die, their shells remain intact and eventually accumulate as sediments on the sea floor. Remarkable for their intricate beauty, these minute skeletons can also yield important clues to oceanographers attempting to reconstruct the earth's climatic history (see Chapter 6).

Far less is known about smaller free-living oceanic amoebae and the marine ciliates and protozoan flagellates. Many of these are in the nanoplankton size range (2 to 20 microns in diameter). Research to date indicates that they are exceedingly numerous and that they probably primarily consume bacteria-sized cells in the picoplankton size fraction. As such they may provide a critical link between the tiniest primary producers in the oceans, including the photosynthetic cyanobacteria, and organisms higher in the food chain.

The Nekton

Asked to list the most familiar marine creatures, most of us would undoubtedly name fish and marine mammals. After all, the finfish, whales, dolphins, and their like are the most conspicuous denizens of the seas and, commercially, the most important to humans. Your brief sojourn in the surface waters of the North Atlantic has, however, turned up few creatures larger than copepods. This is no coincidence. For all their familiarity and inherent appeal, the large swimming organisms comprising the nekton make up only about 0.1 percent of the entire weight of organisms inhabiting the oceans. Still, they are worth spending some time with because they are at the top of oceanic food chains, because they are an important source of human food, and, simply, because they are interesting.

SQUID. Most nekton are vertebrates, but there are some notable exceptions. Wait patiently, and carefully watch the plankton-rich wa-

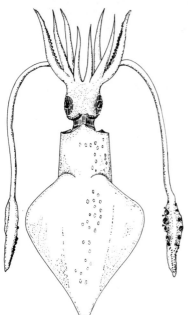

FIGURE 9.27. Common European squid. (Redrawn from *Animals,* a pictorial archive of copyright-free illustrations from nineteenth-century sources selected by J. Harter.)

ters around you. Suddenly, a streamlined form rockets into view, trailing a streamer of tentacles behind. This is no fish, but a mollusk more closely related to a clam than to a tuna. What you have observed is a squid (Figure 9.27), close cousin to the octopus and a diminutive relative of the large deep-water forms described in Chapter 10. A squid is a rapacious predator, seizing a prey organism with sucker-rimmed tentacles, and tearing and rasping its flesh with a parrotlike beak at the base of the tentacles. Squids are literally jet-propelled; water circulates through a collarlike device near the head (the *mantle*) and is explosively expelled through a siphon when the mantle is rapidly closed. The sophisticated neuromuscular coordination required for such motility is enabled by a complex nervous system and surprisingly large brain—certainly remarkable adaptations for an animal whose closest terrestrial cousin is a snail. When one considers that squids also have evolved a highly effective escape mechanism in the form of a characteristic ink cloud, it is easy to appreciate the extraordinary adaptability of these organisms to their pelagic environment. Small wonder that they are among the most abundant nekton in the seas.

BONY FISH. Our nekton watch turns up several squid, but most of our sightings are of vertebrates—notably the finfish. In these waters we are most likely to view representatives of the *bony fish* (class Osteichthyes), which include the familiar herring, mackerel, and tuna. Representative bony fish are shown in Figure 9.28.

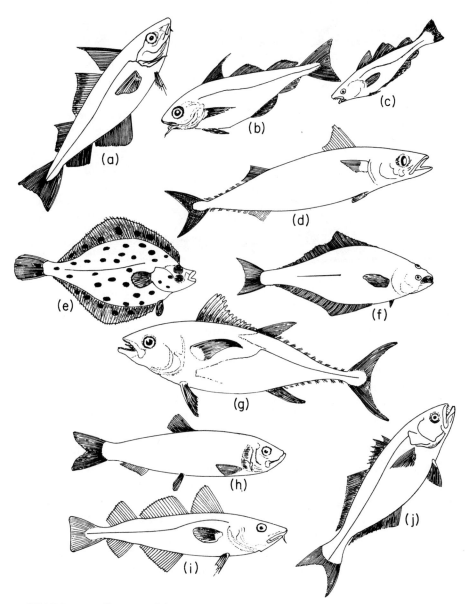

FIGURE 9.28. Common fish. (a) and (b) Haddock. (c) Whiting. (d) Mackerel. (e) Plaice. (f) Halibut. (g) Tuna. (h) Herring. (i) Cod. (j) Bluefish. (Redrawn from *Animals,* a pictorial archive of copyright-free illustrations from nineteenth-century sources selected by J. Harter.)

FIGURE 9.29. Fish school. (Photograph by James Sears.)

Most of the fish we see are swimming in tightly massed and seemingly leaderless aggregations called *schools* (Figure 9.29). Schooling is widespread, though not universal, among finfish and is a fascinating example of well-developed social behavior in the animal kingdom. The size of fish schools varies tremendously. In some species only a few individuals swim together at a given time while, among the herring, single schools may spread over several square kilometers. Characteristically, individual fish are all of the same species, of remarkably uniform size, and occupying constant spacing from each other. This uniformity is maintained even when different schools come together—the individuals sort themselves out. Schools are not strictly leaderless. In fact, the leader constantly changes as the school changes direction.

There are probably several advantages to schooling. Mating and subsequent fertilization of eggs is facilitated by dense aggregations of individuals. Juvenile fish seem to learn better in groups than under isolated circumstances. There is also strong evidence that schooling may confer significant protection against predators. This may seem illogical when we consider that, in the absence of schooling, humans would never be able to efficiently harvest food fish; however, we are the only marine predators which harvest more prey than can be consumed at one time. Other ocean carnivores actually have less chance of decimating a fish population if the latter are bunched into a few widespread groups than if they are uniformly distributed in the environment. In the first place, encounters between predator or prey are less likely than if fish are more widely dispersed. Second, when meetings do occur, predation is ultimately limited by the feeding capacity of the predator. Finally, there is some evidence that the potential predator is confused by the sheer numbers of fish in a school and has difficulty "zeroing in" on an individual fish.

OTHER NEKTON. There are other well-known forms of nekton, but you are even less likely to encounter them because of their comparative scarcity. Perhaps this is just as well in the case of members of the Chrondrichthyes class. These are relatively primitive fish which lack gill covers and bony skeletons; their internal support is provided by cartilage. They include the sharks, skates, and rays. Even more primitive fish include the jawless Agnatha class. Most of them are scavengers or parasites. They are exemplified by the notorious sea lamprey, a rather unattractive creature which makes its living by attaching itself, with its suction-cup mouth and small rasping teeth, to the body of a larger fish and feeding parasitically on its host's fluids.

Reptiles may contribute to the nekton population in some marine areas. Air-breathing sea turtles and snakes are relatively common examples. Finally, marine mammals, including whales, dolphins, porpoises, seals, sea lions, walruses, and sea otters, are well-known nekton. Several of these organisms are discussed in more detail later in this chapter and in Chapter 12.

LOCOMOTION AND MIGRATIONS OF NEKTON. Most of the large marine nekton are capable of rapid, efficient swimming over long distances. Have a good look at a mackerel as it swims into view. Clearly, its streamlined body is built for rapid propulsion. Its tapering form and rounded cross section reduce drag and turbulence, allowing maximum speed with minimal expenditure of energy (Figure 9.30). The actual thrust is provided primarily by muscular movement of the *caudal fin* (Figure 9.31). The shape of caudal fins varies considerably among fish (Figure 9.31) and usually represents a compromise between forms enhancing speed and those aiding maneuverability.

Many of the fish here in these waters do not spend their entire lives in this area but undergo regular migrations between inshore waters and deeper offshore waters. There are significant ad-

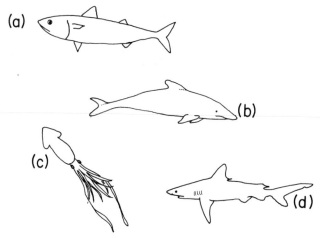

FIGURE 9.30. Streamlined nekton. (a) Herring. (b) Dolphin. (c) Squid. (d) Shark.

vantages to such a nomadic existence. To begin with consider the precarious predicament of a young fish. Predators, often including its own parents, loom hungrily all around. Inopportune water movements sweep it into unfavorable environments. There simply isn't enough food to share equitably with all members of the population. Small wonder that few fry ever make it to adulthood. The high rate of juvenile mortality among pelagic fish is exemplified by the European plaice, of which 99.995 percent die in the first year of life.

To some extent, high mortality is offset by high fecundity. Single spawnings which produce millions of fertilized eggs are not

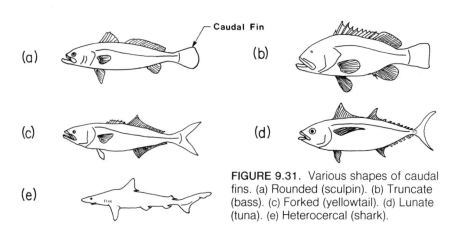

FIGURE 9.31. Various shapes of caudal fins. (a) Rounded (sculpin). (b) Truncate (bass). (c) Forked (yellowtail). (d) Lunate (tuna). (e) Heterocercal (shark).

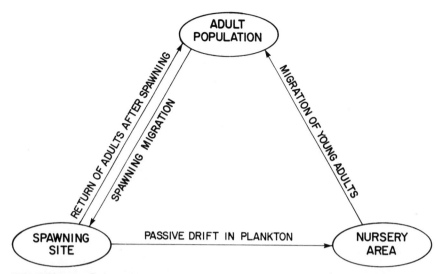

FIGURE 9.32. Generalized migration circuit of migratory nekton. (From Barnes, R. S. K., and Hughes, R. N. 1982. *An introduction to marine ecology, 2nd Ed.* Oxford: Blackwell Scientific Publications.)

uncommon. If each female in a population produces 5 million eggs during the course of a lifetime, 99.99996 percent could die before maturity and still the population would remain constant. Conversely, an exceedingly small variability in survival rate from successive annual spawnings will cause enormous fluctuations in the numbers of fish "recruited" into the population from each spawning year. Consequently, a single highly successful *year-class* (representing the surviving recruits from a single year's spawning) may dominate a fish stock for several years; a series of poor year-classes may bring a stock to the brink of extinction.

Migratory behavior improves the odds of individual fish survival by enabling different developmental stages to inhabit waters which are most favorable for a particular stage. Characteristically, shallow, protected areas serve as nurseries for juveniles, who take advantage of the shelter afforded by benthic plants and the rich supply of planktonic food. Larger fish usually need different food sources; thus, once they reach a certain size, they migrate into deeper waters. Once reproductive maturity is reached, spawning typically takes place in a third region which is situated such that the drifting larvae will be passively carried downcurrent to the nursery area. This triangular *migration circuit* is depicted in Figure 9.32.

Migrations are not restricted to fish; other nekton, including shrimps and squids, may practice them. Occasionally, where re-

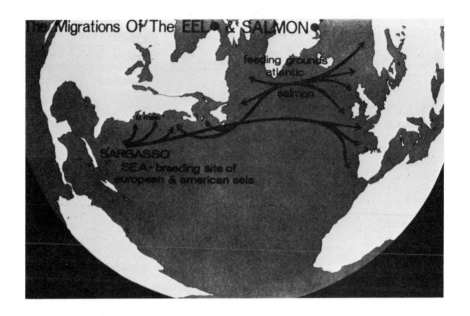

The Migrations Of The EEL & SALMON

feeding grounds
atlantic
salmon

SARGASSO
SEA - breeding site of
european & american eels

stricted circulation makes it unlikely that larvae will be swept into unsuitable waters or when offspring are large and few (e.g., the marine mammals), spawning and nursery areas may be in the same coastal area.

Migrations may cover remarkably large distances. Herring often undertake round trips of 3000 kilometers. The gray whale regularly travels 18,000 kilometers. Even marine reptiles may undergo extensive migrations. A species of green turtle periodically makes a trip from Brazil to tiny Ascension Island—2300 kilometers offshore—to mate and lay eggs (see Chapter 12).

The most familiar migrations are probably those of the salmon. When these *anadromous* species (i.e., living mainly in the sea but returning to fresh water to reproduce) are ready to breed, they return to the same streams in which they were spawned, finding these streams by means of "memorized" olfactory cues.

Examples of well-known oceanic nekton migrations are depicted in Figure 9.33.

INTERACTIONS OF MARINE ORGANISMS IN PELAGIC SURFACE WATERS

To fully understand the interactions of marine organisms it is necessary to review some basic ecological terms and principles. An *ecosystem* is the composite of living (*biotic*) and nonliving (*abiotic*) components in a distinct environment. The sum total of organisms inhabiting a specific ecosystem constitutes a *community*. A *niche* is the ecological "role" and position of a population of one species of organisms within an ecosystem and essentially defines the specific combination of biotic and abiotic requirements and behavioral characteristics which are unique to that species in that ecosystem. It is considered axiomatic that only one species can occupy the same niche in the same ecosystem. If more than one species attempts to occupy the same niche, the ensuing competition will eliminate all but one species.

With this background, let's examine the interrelationships of life forms in the surface waters of the pelagic zone. Consider the various species you have encountered here in the North Atlantic. They differ from one another in a number of ways, including size, lifestyle, and mode of nutrition, yet they are clearly interdependent. Mutually beneficial interactions in marine communities are manifested in a fascinating variety of ways. Large sessile organisms provide support and protection for smaller forms. Many species produce and excrete chemicals, including antibiotics, required by other species. Certain fish and crustaceans even obtain their nutrition by cleaning debris from larger animals. One-celled algae inhabit the tissues of a number of marine animals. The animals provide nutrients and protection for the algae; the algae supply oxygen and, occasionally, nutrition to the animals.

Marine Food Webs

The most important interactions between marine organisms are nutritional. All communities depend on primary producers as their ultimate source of food. In the seas, nearly all primary production is carried out by phytoplankton, since larger, attached benthic plants are restricted to the coastal fringe where light can penetrate to the bottom. This primary production constitutes an enormous reservoir of chemical energy available to higher organisms in the oceans.

Classically, the transfer of energy from lower to higher organisms is represented as a simple *food chain*. In the seas, such a food chain might be represented by zooplankton feeding on phytoplankton, small nekton feeding on zooplankton, and still larger nekton feeding on small nekton (Figure 9.34). Each feeding level is termed a *trophic level*.

Though nutritional relationships in the oceans are usually complex, it is instructive to consider the consequences of even sim-

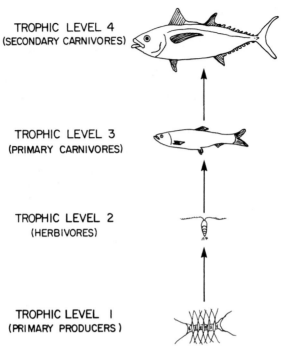

TROPHIC LEVEL 4
(SECONDARY CARNIVORES)

TROPHIC LEVEL 3
(PRIMARY CARNIVORES)

TROPHIC LEVEL 2
(HERBIVORES)

TROPHIC LEVEL I
(PRIMARY PRODUCERS)

FIGURE 9.34. Simple oceanic food chain.

ple food chains. The vast numbers and rapid reproductive rates of phytoplankton in some surface waters might indicate support for a large number of links in the food chain and an enormous quantity of fish in the sea. In fact, this is far from the truth. There are three major reasons.

First, only the upper layer of pelagic waters receives sufficient light for primary production; that production must ultimately supply the nutritional needs of the remaining 98 percent of the dark ocean environment. Second, most of the seas' illuminated regions are deficient in nutrients, thereby severely limiting primary production. Finally, only a fraction of the food energy at any trophic level is converted into energy at the next trophic level. A general rule of thumb is that about 90 percent of available energy is "lost" during transfer between levels; thus, the *ecological efficiency* of a transfer (defined as the food energy actually incorporated into a higher level divided by the energy available to it from a lower level) approximates 10 percent (perhaps as high as 20 percent in some marine environments). Of the nine tenths of the energy which is not transferred between trophic levels, some is not eaten, some of what is eaten is not assimilated, and some is channeled into nonnutritional (*nontrophic*) growth like bones and teeth. A large portion of the

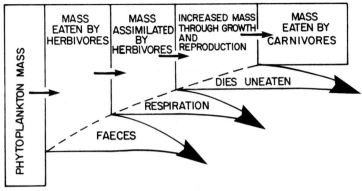

FIGURE 9.35. Energy transfer from trophic level one to trophic level three. (From Thurman, H. V. 1981. *Introductory oceanography.* 3rd ed. Columbus, Ohio: Merrill.)

assimilated energy is "lost" through *respiration*, the process of obtaining energy required for cellular activities through oxidation of organic molecules by cells. The passage of energy through trophic levels is depicted in Figure 9.35.

The consequences of such inefficiency are profound. It means, for example, that 1000 units of phytoplankton food energy will support only one unit of food energy at the fourth trophic level (Figure 9.36). It is not difficult to see why food chains can never be very long and that production of high-level carnivores is decidedly finite.

FIGURE 9.36. Generalized depiction of energy "losses" due to inefficiency of energy transfer between trophic levels. Although transfer efficiencies are probably quite variable, it is assumed here that there is a 2 percent transfer efficiency between radiant energy and primary producers and a 10 percent transfer efficiency between successive trophic levels in an ecosystem. (From Thurman, H. V. 1981. *Introductory oceanography.* 3rd ed. Columbus, Ohio: Merrill.)

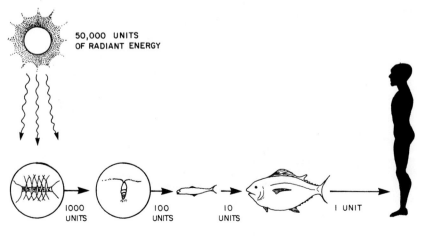

Nutritional relationships between marine organisms are more accurately represented by *food webs* than by food chains. In non-linear food webs many organisms may consume more than one species of prey item and, in turn, may be appealing to more than one species of predator. These options provide resiliency to a community: If a particular species is eliminated through environmental disturbance, the remaining species can pursue alternative feeding pathways. A relatively simple oceanic food web is depicted in Figure 9.37.

It must be emphasized that our understanding of marine food webs is currently undergoing revision as we become increasingly aware and appreciative of the critical roles apparently played by picoplankton (including photosynthetic and nonphotosynthetic bacteria) and microzooplankton. The importance and hypothesized functional roles of these organisms were described earlier in this chapter.

Biological Productivity in the Oceans

Our observations in the surface waters of the North Atlantic have revealed something of the spectacular diversity of the life forms which inhabit the oceans. We have been introduced to their lifestyles, their problems, and their interrelationships. We have a sense of their abundance and fecundity, their exponential proliferation in

FIGURE 9.37. Representative marine food web in the North Sea. (From Thurman, H. V. 1981. *Introductory oceanography.* 3rd ed. Columbus, Ohio: Merrill.)

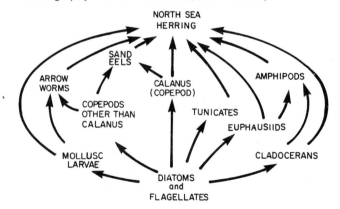

these seasonally fertile waters. We wonder: Are their populations and production representative of the oceans in general? The question is a good one. Knowledge of the numbers and reproductive rates of the seas' plants and animals will not only yield greater understanding of marine ecology but will also give us a better idea of the potential of the oceans to supply food to a hungry world.

In the early days of marine science, most observations were limited to accessible areas in close proximity to laboratories. Primarily, these were coastal waters in temperate latitudes. Researchers quite naturally assumed that these waters were representative of the seas as a whole. These assumptions were short-lived. Once oceanographers could get to sea on a regular basis, it quickly became clear that the productivity of the oceans is enormously variable and the distributions of organisms exceedingly *patchy*. Let's look at the causes of this variability and patchiness and review what we now know about global oceanic productivity. Let's first, however, be sure we understand the concept of productivity.

Supposing that, tired of suburban life, you decide to move to the country and operate a Christmas tree farm. You locate a suitable piece of property, purchase some seedlings, plant them in ordered, uniform rows, and sit back and wait for the profits to roll in. Being a practical sort, you have worked out a careful budget which allows you to project your returns. Critical to this budget is your harvest projection. How did you determine this? To estimate harvest you needed two pieces of information: the total number of trees planted and the speed at which they will grow to maturity.

Estimating productivity in the oceans is a little like working out a budget for a Christmas tree farm. To begin with you need to know the quantity of individuals of a particular species inhabiting a given area at a specific time. This value constitutes the *standing crop* for that species and is usually measured in numbers or *biomass* (weight of living organisms) per unit area (e.g., No./m^3 or gm/m^3). However, standing crop is highly variable, particularly for short-lived, rapidly reproducing organisms like phytoplankton.

To calculate *productivity* you must also assess the *rate* of growth (or of reproduction, in the case of small, short-lived species), taking care to include a period of time long enough to dampen out seasonal variations. Productivity is characteristically measured as units of energy, numbers, or biomass per unit area per unit time (e.g., gm/m^3/year). It is important to note the distinction between *gross productivity* and *net productivity*. In photosynthetic primary producers, the former represents the total rate of photosynthesis and, by extension, production of food energy of organic molecules. The latter refers to the *storage* rate for organic compounds manufactured by photosynthesis after subtracting the rate of respiration (which consumes organic molecules even as they are being manufactured). Net (primary) productivity is what is manifested as actual growth. Similarly, net productivity in heterotrophs represents energy assimilated less energy respired.

Productivity is more easily determined for Christmas trees than it is for oceanic species like phytoplankton. In the former case, one has only to measure annual increases in tree height (and know something about the relationship between height and weight). In the case of phytoplankton, where reproduction is rapid and cells are usually dying or being consumed as fast as they are being produced, productivity cannot be accurately estimated by measuring changes in standing crop over time. Field observations have to be coupled with laboratory assessments of reproductive rates and/or photosynthetic rates.

Variations in Oceanic Productivity

Recall that light intensity and nutrient concentration are the two factors which, more than any other, govern primary productivity in the seas and, by extension, the production of all life therein. Neither factor is uniformly distributed in marine waters, either on a geographical or seasonal basis. The effects of this unequal distribution are readily apparent in comparing the productivities of different marine environments.

With few exceptions, coastal waters are far more productive than offshore waters, even within the same latitudes. This is strictly a function of the greater availability of nutrients in coastal regions, in part because of runoff from land and in part because the shallow waters mix thoroughly. The general paucity of nutrients in the global ocean has led to the description of the open sea as a "biological desert," although, as indicated above, there is growing evidence, from studies of nanoplankton and picoplankton, that offshore waters may be more productive than was believed two decades ago. Exceptions to the comparatively impoverished condition of the open ocean occur where deep, nutrient-rich waters flow up to the surface in upwelling areas far from land. Equatorial upwelling regions provide an example. The ocean's highest productivities occur in coastal upwelling regions, such as that off the coast of Peru, described in Chapters 7 and 12.

Seasons in the Seas

Apart from the contrast between coastal and offshore waters, there are considerable latitudinal and seasonal variations in marine productivity. Ultimately, these variations are based on differences in values and distribution of three environmental factors: light intensity, water temperature, and nutrients. Let's examine how these factors affect annual productivity in polar, temperate, and tropical seas.

If you were to carefully sample the uppermost waters in many oceanic regions in early winter, you would likely turn up a respectable, though probably scanty, array of phytoplankton. Even under the harshest conditions a small population inevitably persists, providing there is a minimum intensity of sunlight penetrating the surface layer. What changes would you see if you were to periodically sample throughout the year? Outside of the tropics, there are pronounced seasonal patterns of light and temperature which are, to a large extent, paralleled by patterns of oceanic productivity. Consider first the temperate latitudes.

PRODUCTIVITY IN TEMPERATE WATERS. For phytoplankton in temperate latitudes in early winter, it is the best of times and the worst of times. Nutrients are plentiful, having been thoroughly mixed into surface waters with the breakdown of the seasonal thermocline in late fall (see Chapter 7). Rather than being inhibitory, the cold temperatures are, indirectly, an aid to production, since they prevent the nutrient-trapping thermocline from reforming. Nonetheless, short daylength and the sun's shallow angle place significant limitations on photosynthesis, and primary productivity is minimal. Low primary productivity also means low populations of consumer organisms.

The responses by plankton to changing seasons in temperate latitudes are highly variable, but we can make some crude generalizations. As days lengthen and light intensity strengthens, phytoplankton populations are dramatically affected. Sometimes as early as mid- to late-winter, but more generally in early spring, a burst of primary productivity occurs, quickly capitalizing on the abundant available nutrients. This sudden production is often referred to as the *spring bloom* or *spring diatom increase* (since diatoms so often

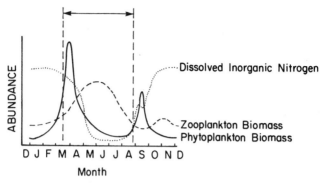

FIGURE 9.38. Generalized seasonal patterns of phytoplankton and zooplankton production in temperate latitudes. Patterns vary considerably between specific locales. (From Barnes, R. S. K., and Huges, R. N. 1982. *An introduction to marine ecology, 2nd Ed.* Oxford: Blackwell Scientific Publications.)

dominate the phytoplankton). Rapid cell-division rates result in geometric increases in population; phytoplankton densities more than 1000 times those found in early winter typically occur.

The riotous population explosion is short-lived. Almost as rapidly as it flowered the bloom fades. There are three principal reasons. First, the plankton quickly deplete their available nutrients. Second, the burst of phytoplankton growth provides a windfall food supply to herbivorous zooplankton. Fattened by their feeding binge, their numbers increase too. The resultant zooplankton *grazing pressure* makes significant inroads on the phytoplankton population. Finally, the thick density of phytoplankton cells has a self-shading effect, so that they ultimately reduce photosynthetic light reaching their numbers even as the sun's intensity increases.

By midsummer phytoplankton numbers are relatively low (though still usually higher than in early winter), and zooplankton populations are in a state of decline as they outstrip their food supply. Excretion by grazers and decomposition of plant and animal remains return some nutrients to the waters, but as surface waters warm, the temperature and density *stratification* established by the developing thermocline prevents the water column mixing which would allow significant nutrient renewal in the photic zone. Summer's end sees a peak in surface-water temperatures, and the waters gradually cool again as fall advances. By late fall, the thermocline again breaks down, nutrients are again mixed into the upper layers, and a brief plankton bloom recurs. The fall phytoplankton increase in temperate waters is smaller and of shorter duration than the spring bloom because declining light intensity effectively nips it in the bud. A generalized depiction of the annual cycle of plankton production in temperate latitude waters is presented in Figure 9.38.

There is one other point to make about the annual plankton cycle in temperate areas. Very often there is a *species succession* associated with the dominant algal species during a phytoplankton bloom. In other words, in the early stages of the bloom one species will predominate, only to soon die out and be replaced by another, and so on. There is some evidence that the release of metabolites by each successive species chemically "conditions" the waters so as to gradually favor the growth of a successor and inhibit its own growth.

SEASONS IN THE POLAR SEAS. We can also make some crude generalizations about productivity in high-latitude waters, recognizing again that considerable variations occur.

There are distinct seasonal variations in productivity in high-latitude waters, but the period of rapid growth and production is extremely attenuated. Surface waters are cold year-round, preventing the formation of a persistent, deep thermocline and its associated stratification of the water column. The well-mixed waters ensure a continuous and ample supply of nutrients in the photic zone. However, these nutrients seldom get used. Except for a brief period in midsummer, the combination of short daylength, a shallow sun's angle, and extensive ice cover severely limits the amount of light available to primary producers.

Ironically, just as polar seas receive the least amount of light in the global oceans through most of the year, they receive the most in midsummer because of the tilt of the earth's axis (Figure 6.3). Ample light and abundant nutrients produce a huge, though short-lived, midsummer phytoplankton bloom which in turn stimulates a large production of zooplankton. A generalized annual cycle of productivity in high-latitude waters is depicted in Figure 9.39.

Included in the abundant midsummer zooplankton of polar seas, particularly in Antarctic waters where they comprise the largest majority of the floating animals, are a subclass of crustaceans called *euphausiids* or *krill* (Figure 9.40). Krill are larger than copepods but slightly resemble them (though they lack the characteristic oarlike appendages). They are at the center of the complex Antarctic food web, which includes penguins and flying birds, whales, seals, squids, and fishes. They are preferentially consumed

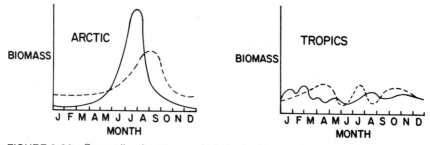

FIGURE 9.39. Generalized patterns of phytoplankton and zooplankton production in the Arctic and the tropics. (From Parsons, T. R., Takahashi, M. and Hargrave, B. 1977. *Biological oceanographic processes*. 2nd ed. Oxford: Pergamon Press.)

by several species of plankton-feeding *baleen whales* which migrate to high-latitude waters during the polar summer. A single baleen whale may ingest up to 850 liters of krill per day.

Such gluttony does not seem to severely tax the krill populations, as evidently there are plenty of euphausiids to go around: By one estimate the annual production of a single species, *Euphausia superba*, is twice the yearly commercial fish harvest in the world. With the decimations of whale stocks because of overharvesting, it is not surprising that some observers have proposed that humans directly harvest the "surplus" protein-rich krill population for their own consumption. In fact, Japan, Poland, Russia, Taiwan, and West Germany have harvested krill for several years, landing nearly 400,-000 metric tons in 1979. However, because of the lack of a firm data

FIGURE 9.40. The krill *Euphausia superba*. (From Tait, R. V. 1981. *Elements of marine ecology*. 3rd ed. London: Butterworths.)

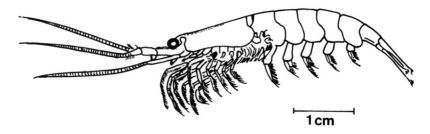

1 cm

base on krill populations, it is difficult to predict what impact extensive commercial krill fishing might have on the Antarctic ecosystem.

PRODUCTIVITY IN TROPICAL SEAS. In contrast to temperate and high-latitude waters, tropical oceans do not experience marked seasonality. This has significant consequences for productivity in these regions.

The most important productivity-related characteristic of tropical waters is the existence of a permanent thermocline because of permanently warm surface waters. Warm water overlying denser, nutrient-rich cold water located below the base of the thermocline restricts mixing of the water column. As a result, the photic zone remains continually deficient in nutrients and, consequently, experiences comparatively low and essentially constant productivity year-round (Figure 9.39).

SUGGESTED READINGS

BARNES, R. S. K., AND HUGHES, R. N. 1982. *An introduction to marine ecology.* Oxford: Blackwell.

BURGESS, J. W., AND SHAW, E. 1979. Development and ecology of fish schooling. *Oceanus* 22:11-17.

CONSIDINE, J. L., AND WINBERRY, J. J. 1978. The green sea turtles of the Cayman Islands. *Oceanus* 21:50-55.

DALE, B., AND YENTSCH, C. M. 1978. Red tide and paralytic shellfish poisoning. *Oceanus* 21:41-49.

DAWES, C. J. 1981. *Marine botany.* New York: Wiley.

GRAY, J. 1957. How fishes swim. In *Oceanography*, pp. 228-34. San Francisco: Freeman. (Readings from *Scientific American.*)

HARDY, A. 1965. *The open sea: its natural history.* Part I: *The world of plankton.* Part II: *Fish and fisheries.* Boston: Houghton Mifflin.

ISAACS, J. D. 1969. The nature of oceanic life. In *Oceanography*, pp. 215-27. San Francisco: Freeman. (Readings from *Scientific American.*)

MACLEISH, W. H. 1978. Marine mammals. *Oceanus* 21(2). (A collection of articles.)

NEWELL, G. E., AND NEWELL, R. C. 1977. *Marine plankton, a practical guide.* London: Hutchinson.

NYBAKKEN, J. W. 1982. *Marine biology, an ecological approach.* New York: Harper and Row.

Russell-Hunter, W. D. 1970. *Aquatic productivity.* New York: Macmillan.

Ryan, P. R., ed. 1982. Sharks. *Oceanus* 24(4). (A collection of articles.)

Shaw, E. 1962. The schooling of fishes. In *Oceanography,* pp. 235–243. San Francisco: Freeman. (Readings from *Scientific American.*)

Sieburth, J. McN. 1979. *Sea microbes.* New York and London: Oxford University Press.

Sumich, J. L. 1984. *An introduction to the biology of marine life.* 3rd ed. Dubuque, Iowa: Brown.

Tait, R. V. 1981. *Elements of marine ecology.* 3rd ed. London: Butterworths.

Chapter Ten

Life in the Sea II:
The Deep Sea

DARK. COLD. CRUSHING PRESSURE. FANTASTIC CREATURES GLID-ing, ghostlike, in eternal silence. An archaic, serpentine monster or two lurking in some remote unplumbed abyss.... What a host of images are evoked by the deep sea.

Are these images accurate? What is the deep sea, and what is it really like? For most of us it is, at best, a vague and hazy concept, one that probably piques our interest but, like intergalactic space, is so far removed from our everyday world that we have relegated it to the realm of fancy or even fantasy. In fact, scientists know surpris-ingly little about the submarine environment. It is a world exceed-ingly difficult to study, where long, expensive labor under the most trying conditions may yield but limited, fragmentary knowledge.

Difficult though it may be, we must study the deep sea. It is the most extensive habitat on earth, comprising two thirds of the earth's living space and 90 percent of the oceans' volume. It is a cru-cial component of the global ecosystem. It regulates and modifies climate. The earth's most fundamental geological processes occur and are recorded in and on its floor. The deep sea harbors a myriad of obscure life forms, many of which are probably unknown to sci-ence. It is an existing or potential source of food, minerals, and en-ergy. It is a receptacle for our most poisonous wastes.

Less than a century and a half ago, there was little incentive to study the deep sea. Its frigid, sunless depths suggested an impos-sibly harsh environment incapable of supporting the most rudimen-tary life forms. In fact, an esteemed, early nineteenth-century natu-ralist, Edward Forbes, concluded that the waters below 600 meters were totally devoid of life (i.e., *azoic*). So popular was this *azoic theory* that it persisted into the middle of the last century, despite infrequent evidence to the contrary. (For example, decades earlier pioneer oceanographers John and James Ross had recovered intact living organisms at depths of well over a mile, but their reports were largely ignored.)

In the latter half of the nineteenth century, two events revolu-tionized existing thought about the deep sea. The first was the pub-lication of Charles Darwin's *The Origin of Species* in 1859. The second was the *Challenger* oceanographic expedition from 1872 to 1876.

In his epochal treatise, Darwin was the first scientist to amass a persuasive body of evidence to support the argument that living or-ganisms undergo gradual evolutionary changes over time in re-sponse to environmental pressures, making it likely that a modern

FIGURE 10.1. H.M.S. *Challenger*. (From Anikouchine, W. A., and Sternberg, R. W. *The world ocean.* Englewood Cliffs, N.J.: Prentice-Hall.)

species will scarcely resemble its early ancestors. The agent of change, Darwin hypothesized, is natural selection of environmentally suitable and heritable characteristics, that is, features that confer successful adaptation to a changing environment. To nineteenth-century oceanographers, the implications were profound. In their assessment, the remote recesses of the deep sea, despite being rigorous, were, of all the globe's environments, the least likely to have undergone changes since the origin of life on earth. Thus, it was reasoned, the deep sea should be a repository for primitive species under no pressure to change over the eons—a living museum for ancient, fundamental life forms whose terrestrial and shallow-water cousins have subsequently evolved beyond recognition. In the formless ooze that blankets the abyssal deeps, some scientists even imagined transitional forms, not quite life and not quite nonlife—in Loren Eiseley's words, "life in the process of becoming."

It was against this background that the *Challenger* expedition was launched in 1872 (Figure 10.1). Directed by Sir Charles Wyville Thomson and staffed by a team of naturalists, the expedition lasted four years and traversed more than 100,000 kilometers of ocean. Thousands of samples were collected, and a staggering amount of data was compiled. Fifty massive volumes were published; even so, some samples remain unanalyzed to this day.

What was learned? That the deep sea is indeed inhabited, but mainly by comparatively modern forms. While some living fossils

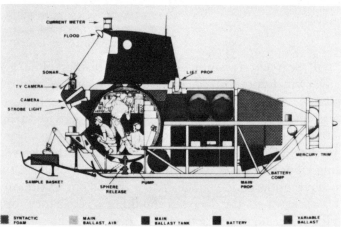

FIGURE 10.2. (a) *Alvin*. (b) Cutaway view of *Alvin*. (Courtesy of Woods Hole Oceanographic Institution).

may be found, most species are of recent origin. Rather than harboring the origin of life, the ocean depths apparently were populated rather late by creatures from shallower waters. These findings disappointed many scientists of the day and devastated some cherished theories, but the expedition had a far-reaching significance, for it marked the birth of oceanography as a true science.

Would you like to know more about the deep sea? Are you interested in sorting out fact and fancy, curious about what we currently know of this vast environment, intrigued by what mysteries remain unfathomed? Then join us in an odyssey to the depths.

A JOURNEY TO THE DEEP SEA

The Research Submersible Alvin

Our vehicle is the manned submersible, R/V *Alvin* (Figure 10.2). Operated by the Woods Hole Oceanographic Institution (WHOI), she is only 7.6 meters long and can barely accommodate a pilot and two scientific observers but, with her titanium hull and frame, can dive to depths of 4000 meters. Built in 1964, she averages over 100 dives a year. With a large stern propeller, two small side-lift propellers, and five viewing ports, she is a highly maneuverable underwater observatory.

At *Alvin*'s forward end a powerful strobe light can illuminate activities and scenes in the gloomiest of underwater environments. To record these scenes and activities, still and TV cameras are mounted adjacent to the strobe light. Protruding from her hull, remotely controlled mechanical arms can retrieve delicate samples and perform complex tasks. Less than a year after her construction, *Alvin* proved her worth when she located and helped recover a U.S. hydrogen bomb accidentally dropped in 800 meters of water off the coast of Spain.

Time to board now. *Alvin* is securely lashed to the side of her "mother ship," the R/V *Atlantis II* (Figure 10.3). Your dive site is

FIGURE 10.3. R/V *Atlantis II*. ((Courtesy of Woods Hole Oceanographic Institution).

the Galapagos Rift, about 200 miles northeast of the Galapagos Islands in the eastern Pacific. The Rift is part of the worldwide system of underwater mountain ranges and oceanic spreading centers which are an integral component of global plate tectonics described in Chapter 4. Since 1977 *Alvin* has made a series of dives to the Rift to study the region's geology and biology. Your task today: to study an unusual community of living organisms associated with an underwater hot spring or volcanic vent in the Rift. The vent and community were originally discovered by WHOI scientists in 1977.

Descent to the Depths

"Close hatch!" The command from *Atlantis II* crackles into the single-sideband radio receiver, startling you out of your reverie. The pilot reaches up and swings down the heavy, watertight door. Twisting the valve wheel shut, he ensures a tight seal. Too late to back out now, and no time to think about claustrophobia. Your mouth dry with anticipation, you listen to the pilot communicate with the mother ship. Five minutes to castoff. A final systems check shows that all is in order. Fortunately, seas are light, and skin divers from *Atlantis II* will have no difficulty releasing the tethers. Overhead, cotton-ball clouds on an azure carpet tumble past the conning tower portholes. Delicate shafts of sunlight fracture and scatter, like shards of golden glass, on the choppy seas. Fix the scene in your mind. You are about to enter a sunless, airless world.

"Cast off!" The divers work quickly, releasing the securing lines. The pilot adjusts *Alvin*'s buoyancy. Slowly at first, you begin to descend. Five meters. Ten meters. Overhead, you can still see sunlight dancing on the waves, but the watery world outside is dimming fast, as if a giant rheostat were gradually being turned down. Suspended material in sea water reflects, scatters, and absorbs incident light. If you were close to shore, where there is an abundance of silt, sediment, and minute life forms in water, the attenuation of light with depth would be particularly pronounced. Over half of the surface light would be lost in the first meter and 99 percent would be extinguished in 15 meters. In contrast, in the clearer waters offshore, like those here in the eastern Pacific, the *99-percent extinction depth* occurs at about 125 meters, two minutes away at your descent speed.

Light is not only dimmed with depth, its color is also altered (Figure 10.4). Imagine a giant palette, displaying the primary hues, attached to the hull outside. Watch the palette closely. By 15 meters

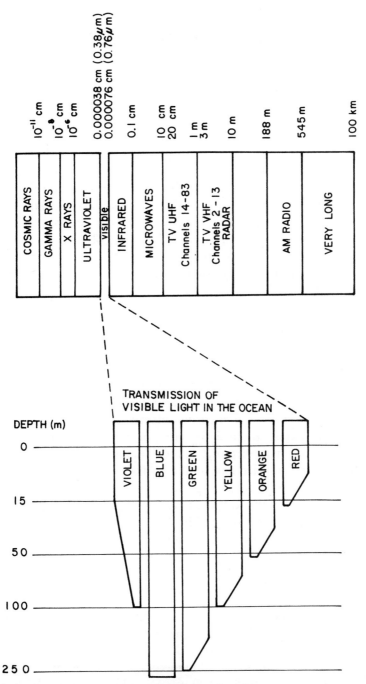

FIGURE 10.4. Electromagnetic spectrum and the transmission of visible light in the ocean. (From Thurman, H. V. 1981. *Introductory oceanography*, 3rd ed. Columbus, Ohio: Merrill.)

the red disappears. The yellow is gone at 100 meters. Blue and green may persist to 200 meters. This selective absorption is based on the fact that visible light is composed of a spectrum of wavelengths. This is readily apparent when light is shone through a prism. The wavelengths are differentially bent; specific wavelengths display specific colors. The shortest wavelength appears as blue, followed by, in order, green, yellow, orange, and red. As light penetrates water, the longest wavelengths are absorbed first and the shortest last. Thus only blue and green colors are visible at depth. A practical analogy will illustrate the point. Are you a skin diver? Pay close attention the next time you cut yourself on a piece of sharp coral far below the water's surface: Your blood will run green.

At 150 meters you are suspended in a cobalt-blue void of unimaginable peace and stillness. Occasional ethereal shapes loom at portholes, then glide off as silently as they arrived. You are at the threshold of the deep sea.

The Bounds of the Deep-Sea Environment

What precisely is the deep sea? While there is no clear distinction between the deep sea and the overlying surface waters, most oceanographers perceive it as the portion of the ocean environment which lies below the depth of visible light penetration and beyond the shallow continental shelves, thus effectively comprising the entire world ocean below about 200 meters of depth. To truly appreciate the vastness of this environment, recall our description in Chapter 3: Half the earth's surface lies 3000 to 6000 meters underwater. If the entire earth's crust, including the continental land masses, were smoothed out, it would be easily swallowed by a global ocean averaging 2440 meters deep. The land's largest mountain, Everest, at 8841 meters, could be deposited in the ocean's deepest trench and still leave over a mile of water above its peak.

Temperature in the Deep Sea

Two hundred meters. You glance at the water-temperature monitor. From a high of 20° C at the surface, the mercury has plunged steeply. This sharp temperature decline, or *thermocline* (see Chapter 7), will persist to about 1000 meters, then level off at about 4° C from there to the bottom, 2500 meters below *Atlantis II*. A salient feature of the deep sea is that it is numbingly, unremittingly cold, with an average temperature only a few degrees above freezing.

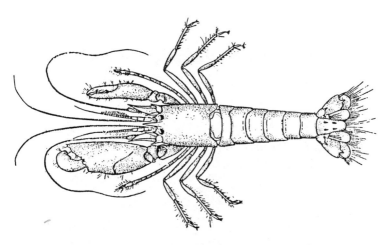

FIGURE 10.5. The snapping shrimp, *Alpheus formosus* Gibbes, ×1.5. (Reprinted with permission of the Vanderbilt Museum, Centerport, N.Y.).

Even in tropical latitudes, the thermocline separates the denser, isothermally cold, deep waters from the comparatively balmy surface waters. Because of their greater density, these deep, cold waters, whose origin is in the surface areas of the polar seas, are trapped below a thin skin of warmer surface water.

Far below the heating influence of the sun, the deep sea displays a remarkably uniform temperature which is essentially unvarying with season or latitude. Such cold presents no major problems for deep-sea organisms, whose metabolisms have adjusted accordingly. In fact, in the unvarying temperature of the depths, deep-sea creatures experience a virtually benign environment compared to their shallow-water cousins, who may alternately bake and freeze on a seasonal, or even daily, basis.

Sound in the Deep Sea

Suddenly, a ghostly moan fills the submersible. With relief you remember the hydrophone mounted outside the hull. It is a popular misconception that the deep sea is a silent world. While it is true that few of its denizens possess vocal cords, at times the depths produce a veritable cacophony of noise. Nearly every imaginable sound has been recorded from the sea, including shrieks, cackles, mews, grunts, and squeals. How are such sounds produced? Apparently in a variety of ways. Fish can resonate their *swim bladders*, gas-filled organs whose primary function is to regulate buoyancy. Snapping shrimp click their claw joints together (Figure 10.5). While one

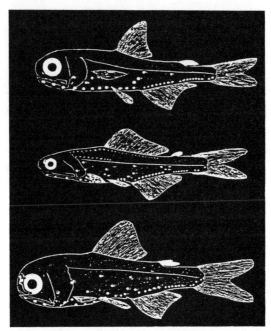

FIGURE 10.6. Large eyes in deep sea lanternfishes. Top: *Myctophum punctatum* (about 10 centimeters long). Middle: *Lampanyctus elongatus* (about 15 centimeters long). Bottom: *Diaphus metopoclampus* (about 4 centimeters long). (From N. B. Marshall. 1954. *Aspects of deep sea biology*. N.Y.: The Philosophical Library.)

shrimp's solo is insignificant, the concert of a congregation can be heard for four kilometers. Lobsters rub their antennae on their shells. Some fish grind together the teeth located in their throats.

Visibility in the Deep Sea

The depth gauge now registers 1000 meters. Utter darkness surrounds the submersible. Time to switch on the strobe light. Almost immediately a fish swims into view, peers curiously into a porthole, then tail-flicks away. Something puzzles you about that fish—something not quite right. The fish is certainly normal-looking for a fish—probably a member of the cod family. That's it—it's too normal. Here, far below the depth of visibility, swims a fish with well-developed eyes, seemingly useless in their stygian surroundings. A chance wanderer from above as out of place as you in this somber world? No. This creature lives here.

In fact, you have made a surprising discovery—that, down to 1.5 kilometers or so, the visual organs of many deep-sea fish remain well developed (Figure 10.6). Some eyes even grow larger and more complex with depth.

The presence of such well-developed eyes in the depths is something of a mystery. There are several possible explanations. We have said that, even in the clearest waters, one percent of the sunlight penetrates only to 125 meters or so. At or below that depth "darkness," for all practical purposes, prevails, but how deep can the tiniest fractions of light be perceived? Physiologists have found that the human eye, after a suitable adjustment period, can detect an intensity only one ten-billionth that of full sunlight. This miniscule amount has been measured at a depth of 550 meters, a full third of a mile below the surface.

The benefit of humans being able to detect such exceedingly dim light is not immediately apparent, but there is reason to believe that deep-sea fish have similarly or even better, developed visual acuity. It is also possible that these visually well-endowed fish make periodic excursions into the illuminated zone far above, perhaps for feeding purposes. Regular vertical migrations are well documented in the deep sea and will be discussed later.

A likely explanation for well-developed eyes in some deep-sea species lies in the phenomenon of *bioluminescence*, the production of light by living organisms (see Chapter 9). While we are perhaps most familiar with the terrestrial example of the firefly, biolumine-scence is widespread in the ocean and well represented in the deep sea. The light is produced chemically in special organs called *pho-tophores*. Since photophores are arranged in patterns which are distinct for individual species (Figure 10.7), it is likely that biolu-minescence functions in species recognition (facilitating location of a mate) and in reproductive behavior. There is also evidence that intense light production is used by some species to distract or blind predators or even illuminate prey.

Below about 1.5 kilometers, most fish and invertebrates have small or degenerate eyes, or visual organs are absent altogether. This presumably necessitates reliance on the other senses—smell, touch, and the *lateral line system* (organs arrayed along the sides of most fish which are used to detect vibratory impulses)—for detection of prey, predators, or mates.

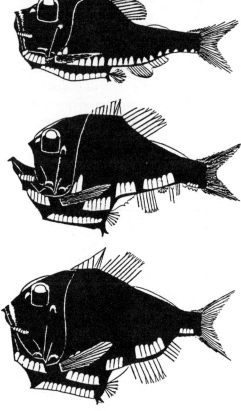

FIGURE 10.7. Different arrangements of photophores in three species of *Argyropelecus*. (From N. B. Marshall. 1954. *Aspects of deep sea biology*. N.Y.: The Philosophical Library.)

Pressure

Two thousand meters. Outside, the ocean presses with a force of nearly one and a half tons on every square inch of *Alvin's* hull. For every 10 meters below the ocean's surface, water pressure increases by one atmosphere, or 1.013×10^6 dynes/cm^2 (14.7 pounds per square inch). The effect of such pressure on compressible structures is formidable. Consider what would happen to an infinitely compressible, half-meter-diameter rubber ball lowered into the sea. At 10 meters, neglecting temperature effects, the diameter would be halved and at 20 meters reduced to a third of its original size. At 2000 meters, your current depth, the ball diameter would be equivalent to the thickness of a nickel. Fortunately, *Alvin's* pressurized, titanium hull has a safe working depth of 4000 meters.

Why are deep-sea organisms not crushed to pulp by the enormous pressures of their environment? For one thing, like all living organisms, they are composed mainly of water, which is nearly incompressible. Nonetheless, there is evidence that high pressures can have deleterious effects on essential cellular processes of unadapted species; deep sea residents have apparently undergone appropriate metabolic adjustments, as yet undefined.

At 2200 meters, the bottom is only a few minutes away. The strobe light stabs into the inky expanse. Something swims lazily into its beam, pausing long enough for you to see it clearly. This is no normal-looking creature, but as bizarre a form as you might encounter in your most vivid dreams. The fish—for it is recognizable as a fish—appears to be all mouth, with large, hinged jaws, long, inward-curving teeth, and a proportionately tiny body. The pilot identifies it as a deep-water angler fish (Figure 10.8). The morphology is characteristic of many deep-sea fish and, despite its unlikely appearance, the fish is superbly suited for the most significant problem faced by organisms in the deep sea—a scarcity of food.

Nutritional Problems in the Deep Sea

In the absence of light in the depths, there can be no photosynthesis, the process which allows green plants to capture light energy and use it to convert nonliving materials into living tissue. Such primary productivity is the supporting base of the food chains in the earth's illuminated or photic environments (see Chapter 9). In contrast, most creatures of the ocean depths, in order to support their metabolic needs, must ultimately rely on chance leftovers which sink from productive surface waters above. Thus, the amount of food available to deep-water consumer organisms will depend on how productive the surface waters are and how deep the consumer is, since the probability of a morsel remaining uneaten, while sinking, decreases with depth.

We have little direct knowledge of the quantity and quality of deep-sea food which originates in surface waters. Most likely it is primarily comprised of slow-degrading, relatively unpalatable remains, such as crustacean shells, vertebrate bones, and terrestrial vegetation. However, even this material can eventually be broken down into a nutritious meal by bacteria. Another apparently important source of deep-sea food is the *fecal pellets* of the myriad zooplankters in surface waters. Remarkably, these pellets are virtually

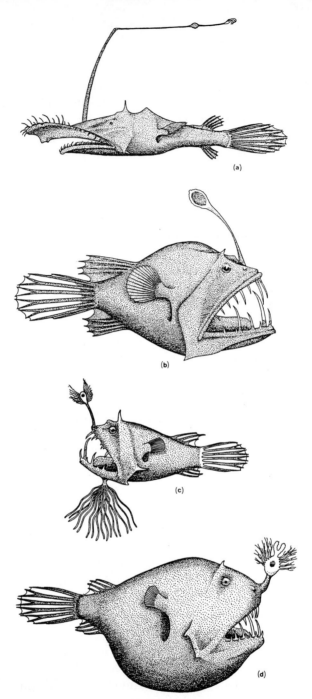

FIGURE 10.8. Deep-sea angler fish. (a) *Lasiognathus saccostoma*. (b) *Acentrophryne longidens*. (c) *Linophryne polypogon*. (d) *Borophryne apogon*. (From the Danish *Dana* expedition, 1920–22. Courtesy of the Carlsberg Foundation, Copenhagen.)

packaged for export to deep waters, being encapsulated in a decomposition-resistant covering and possessing a shape and density which facilitates sinking. Further, there is every likelihood that rapidly descending remains of large fish and mammals may reach the depths quickly enough to provide a relatively intact food source. Finally, *"marine snow,"* consisting of dissolved or colloidal organic matter and gelatinous remains of plankton, may constitute an important food source.

It is one thing for uneaten food to get to the deep sea and quite another for the abyssal residents to find it in the permanent darkness. The preliminary evidence is that they find it rapidly. Kenneth Isaacs and Robert Hessler of the Scripps Institution of Oceanography have found that bait placed on the deep-sea floor is almost immediately detected by organisms from great distances. Presumably, detection is by olfactory means.

Pelagic deep-sea fish must immediately seize and ingest living or dead food of a wide range of sizes as it sinks or swims by. Food encounters in the depths are simply too chancy to waste, and predators cannot afford to be selective. Hence the bizarre morphology of many deep-water fish, exemplified by the angler fish and the gulpers and vipers. Their large jaws and distendable stomachs allow them to seize and swallow prey often twice their size (Figure 10.9). To appreciate this, imagine yourself attempting to swallow a 300-pound steer in one gulp. Since swimming with such a load would be cumbersome at best, digestion presumably is rapid.

Not content to wait for food to happen by, some deep-sea fish, including the anglers and stomiatoids, actually lure their prey with photophores on their dorsal fins or chin barbels, or even in their gaping mouths.

Diurnal Vertical Migration

One of the oldest unsolved puzzles in oceanography may be related to adaptive responses by deep-sea organisms to the scarcity of food in their environment. The puzzle is the phenomenon of *diurnal vertical migration*, the daily up-and-down movement of planktonic and swimming creatures in the water column (Figure 10.10). First discovered by scientists on the *Challenger* expedition, diurnal vertical migration (DVM) is practiced by many deep-sea and middepth organisms, including zooplankton, jellyfish, worms, prawns, shrimps, squids, and fish. The migration is regular and reasonably

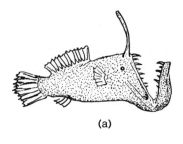

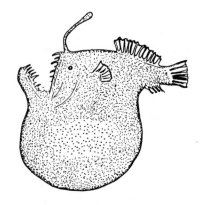

(a)

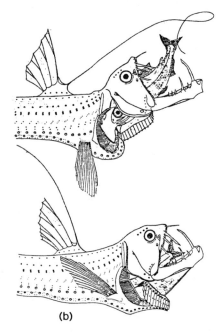

FIGURE 10.9. (a) Deep-sea angler fish before and after swallowing prey. (From Hardy. A. *The open sea*. 1965. Boston: Houghton Mifflin.) (b) *Chauliodus* swallowing a lantern-fish, showing hinging of jaws. (From N. B. Marshall. 1954. *Aspects of deep sea biology*. N.Y.: The Philosophical Library.)

(b)

predictable. As dusk approaches, organisms swim toward the surface, moving higher as light gets progressively dimmer. As dawn arrives, they sink again. The length of the journey may be considerable; some fish and crustaceans move through a vertical distance of 400 meters and subject themselves to temperature changes of 15 ° C and pressure differences of 40 atmospheres.

While it seems evident that animals normally living near the surface are probably responding to changing light intensity, some migrators remain 3 to 5 kilometers below the surface at all times, far below even the dimmest light. This suggests that innate biological rhythms may also be involved.

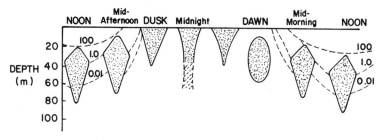

FIGURE 10.10. Generalized "kite" diagram of diurnal vertical migration. Shaded areas show vertical distribution of epipelagic zooplankton at different depths. Dashed lines represent approximately equal light intensity. (From Tait, R. V. 1981. *Elements of marine ecology*. 3rd ed. London: Butterworths.)

    Swimming upward a half a kilometer or more on a daily basis is clearly a large expenditure of energy for a creature of any size. For the trait to persist in a population there must be a net advantage to the organism. What is the advantage?

    The answer is not immediately forthcoming. Obviously, there is more food in the surface layers, so upward migration is reasonable. Deeper migrating organisms, which never make it into surface waters, are presumably following migrating prey above them. But then why bother to move back down into food-deficient waters? Some scientists think that a return to a dimmer environment during daylight hours is a mechanism to escape detection by predators; however, since many descending organisms advertise their presence with glowing bioluminescence, this hypothesis seems questionable.

    It has also been suggested that DVM is an energy-optimizing device: Migrators make short-term feeding forays into warm surface waters but, after feeding, return to deep, cold waters where metabolic rates are reduced. Sir Alister Hardy, Britain's eminent biological oceanographer, has proposed another solution: that DVM extends the horizontal feeding range of migrating organisms by allowing weak swimmers to encounter a range of currents which can transport them into different water masses with new sources of food.

Other Adaptations
of Deep-Sea Organisms

Despite food-maximizing adaptations, large meals are few and far between in the deep sea; consequently, one might expect its residents to be correspondingly small. This is largely true of the bony fish. Seldom more than a few centimeters long, their large mouths

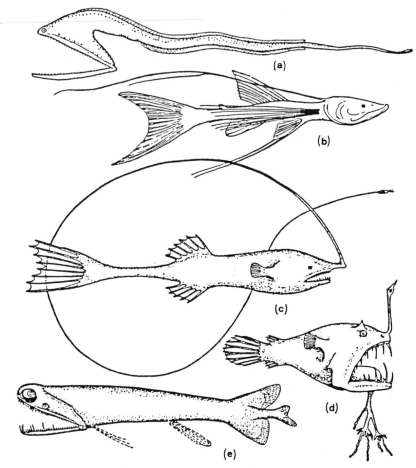

FIGURE 10.11. Several species of deep sea fish, drawn approximately to scale. The top fish is about 15 centimeters. (From Sverdrup, H. U., Johnson, M. W., and Fleming, R. H. 1942. *The oceans:* Englewood Cliffs, N.J.: Prentice-Hall.)

and small bodies maximize food-capture ability relative to body mass which must be supported. Representative deep-sea fish are shown in Figure 10.11. In marked contrast, some invertebrates of the sea floor tend to be quite large—often much larger than their shallow-water relatives. In view of the scarcity of deep-sea food, we don't know why this is so, though there is some evidence that these organisms are long-lived. Perhaps great longevity compensates for slow growth.

The scarcity of food in deep waters also means that the abundance of individuals within any species is likely to be low and that it may be difficult to find a mate. The angler fish *Linophryne argyresca* resolves this problem in a unique way. The females are comparatively large, but the males are tiny and, once they successfully encounter a mate, they remain permanently attached to her, living parasitically on her body via interconnected bloodstreams. His sole function is to provide sperm for reproduction and, as such, he is little more than a reproductive organ on the female (Figure 10.12).

While large size is difficult to support in the food-limited environment of the deep sea, nautical lore is replete with reports of deep-sea monsters. Do such tales have a basis in fact? If, by monsters, we mean huge, uncommon creatures whose size rivals that of the whales, then the answer is yes. The existence of giant squids, whose length may exceed 15 meters (two thirds of which is tentacles) is well documented. The bizarre-looking oarfish, whose head is faintly equine, may be equally large and may be the source for many monster stories (Figure 10.13). The recent discovery of an eel larva nearly 2 meters long suggests that the depths may harbor adult eels ten times that size. What about the fabled sea serpents? While most reports of these creatures are attributable to mistaken identity

FIGURE 10.12. A female angler fish, *Linophryne argyresca,* with a permanently attached parasitic male on her underbelly. While the female is less than 10 centimeters long, she is huge in comparison with her mate. (From the Danish Dana expeditions, 1920–22. Courtesy of the Carlsberg Foundation, Copenhagen.)

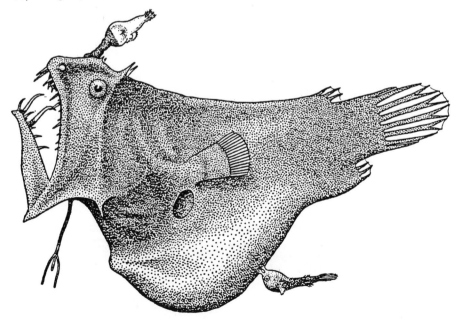

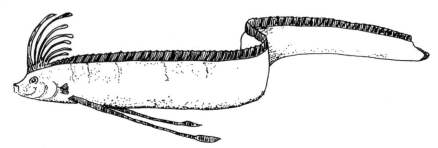

FIGURE 10.13. The oarfish, *Regalecus glesne*. (Redrawn from the Century Dictionary and Cyclopedia, 1904.)

of more common ocean denizens, it has been suggested that whale-sized, fish-eating plesiosaurs (Figure 10.14), believed to have perished with the dinosaurs, may yet inhabit the depths of oceans and deep lakes with free connection to the sea.

A VISIT TO THE GALAPAGOS RIDGE

Twenty-five hundred meters. Suddenly the sea floor looms below, illuminated by the wash of the strobe light. The terrain is rugged, more rugged than any you have seen on dry land. Jagged peaks rise

FIGURE 10.14. Representation of an ancient plesiosaur. (From Hardy, A. 1965. *The open sea.* Boston: Houghton Mifflin.)

in seemingly endless procession, interrupted by deep chasms and smooth shapeless mounds of volcanic rock. You are hovering above a section of the Galapagos Ridge, part of the worldwide system of underwater mountain ranges, or *oceanic ridges,* which mark the boundary of two separating tectonic plates on the earth's surface (Figure 10.15). A deep valley, or rift, cleaves the crest of the Ridge. Liquid lava, erupting from the bowels of the earth, gurgles into fissures in the rift. On opposite sides, the sea floor slides imperceptibly away, driven by unseen, ill-defined forces.

An oceanic ridge and rift system constitutes a dynamic world of earth-wrenching violence and molten beauty, of cataclysmic power and instability. It is the last place, one would imagine, where delicate life forms could survive and even thrive. Or could they?

FIGURE 10.15. Location of the Galapagos Ridge in the East Pacific Rise. (Courtesy of the Woods Hole Oceanographic Institution.)

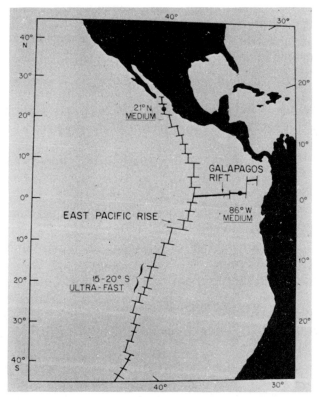

It is widely assumed that there is generally a low abundance of life in the deep sea because of the scarcity of food. In fact, we have few direct observations to support this assumption. It is so extraordinarily difficult to adequately and representatively sample the swimming and drifting denizens of the vast abyssal environment that we remain woefully ignorant of the numbers, species composition, diets, metabolism, and reproductive patterns of the pelagic fish and invertebrates throughout most of the ocean depths.

We know more about the benthic animals, simply because they are relatively stationary and more amenable to capture. Much of our knowledge of the creatures living in or on the sea floor has occurred since the early 1960s, with the development of sophisticated sampling devices and a major intensification of sampling effort. The results have been surprising.

We have learned that the number of animals on the deep-sea floor does indeed appear small and apparently diminishes with depth. However, there is considerable evidence that there is a large number of deep-sea species. In fact, deep-sea faunal diversity may well be among the highest for any habitat on earth. Representative species encompass most of the major phyla encountered in the oceans and include sponges, jellyfish, comb jellies, a variety of worms, bryozoans, sea lilies, starfish, brittle stars, sea urchins, sea cucumbers, sea spiders, snails, slugs, clams, squids, octopuses, barnacles, shrimps and other shrimplike crustaceans, crabs, tunicates, eels, and fish.

High diversity in any living community generally indicates a minimum of competition between species since, when more than one species competes for a scarce resource, the best adapted will characteristically exclude the others. How can competition be minimized in the deep sea when food is so scarce?

Perhaps, in its monotonous constancy, the environment is sufficiently stable and predictable to allow long-term *biological accommodation* between species. For example, the development of highly specialized feeding strategies, which are species-specific, might circumvent direct conflict between species, allowing them to coexist. Thus, stability of the environment is believed to permit diversity among life forms.

However, the evidence is that, instead of being specialists, most deep-sea-bottom dwellers are generalized scavengers, nonselectively ingesting anything and everything of potential nutritional value on the sea floor. Some of these animals are active predators, but most are *infaunal deposit feeders*, which means they live in the sediments and engulf sedimentary material whence they obtain their nutrition. Still other benthic feeding types include the *epifaunal suspension feeders*, which live on the bottom and filter fine food particles from the water. Finally, a number of deep-sea animals, known as *croppers*, appear to combine predation and deposit feeding by indiscriminately scooping up deposit feeders and sediment from the sea floor.

The nonselective feeding strategies employed by croppers in particular apparently result in intensive deep-sea predation. Thus, some researchers consider it unlikely that any one species can become sufficiently abundant to compete directly with another for food or space. According to this hypothesis, many species can live together. Whatever the explanation, the general picture, thus far, is of numerous species but small populations on the deep-sea floor.

A Deep Sea Thermal Vent Community

In 1977, oceanographers discovered an astonishing exception to the general picture of impoverished deep-sea populations. Would it surprise you to learn that, in the catastrophic environment of underwater spreading centers, some of the most productive animal communities on earth thrive in riotous profusion? Recent discoveries have revealed that some rift valleys harbor teeming oases of life, scattered here and there like lush islands in a barren sea. The animal communities are associated with hot-water geysers, or volcanic thermal vents, which erupt through fissures in the lava—vents like the one you will study today, the one which *Alvin* is searching for now.

Alvin's search is facilitated by detailed knowledge and maps of the bottom topography of the region, and by the submersible's horizontal scanning sonar system. The maps have been constructed from photographic surveys carried out with ANGUS (Acoustically Navigated Geophysical Underwater System), a remotely operated, unmanned sled, equipped with cameras and lights, which is towed above the bottom by a surface vessel. Another deep-sea camera system is shown in Figure 10.16.

"There it is!" Illuminated by *Alvin*'s lights, a plume of turbid water jets out of a 25-foot "chimney" of black lava. When the

FIGURE 10.16. Underwater camera system.
(Courtesy of Woods Hole Oceanographic Institution.
Photo by Tom Bolmer.)

350° C plume hits the near-freezing sea water, a fine ash of heavy metals precipitates out and rains down at the base of the chimney. A few feet from the plume, a flash of red color draws your attention. *Alvin* moves in closer, homing in with the TV camera and video-taping the scene for later study. The color, visible because of *Alvin*'s powerful floodlights, proves to be the feathery red tips of a grove of giant worms weaving sinuously from cementlike tubes anchored to the sea floor. A profusion of albino crabs and eel-shaped fish scurries about the worm tubes, and masses of foot-long clams lie tumbled about nearby. Earlier studies have shown that this vent area harbors a spectacular community of animals, some four times as dense as the most productive surface waters (Figure 10.17). *Alvin*'s probe registers temperatures of 15 to 20° C in the waters of the community, near-tropical conditions compared with those of the surrounding waters.

Abnormally warm temperatures are not sufficient to explain the biological extravagance of the vent community; an inviolable law of nature is that populations are irrevocably held in check by the availability of food. What could be the ultimate source of food in these isolated deep-water ecosystems, so removed from the photo-synthesis-stimulating rays of the sun which fuel the primary productivity of green plants, basis of the terrestrial and shallow-water food chains?

The answer is that primary productivity of a different sort alto-gether is occurring here. Vent waters spew forth enormous concen-trations of the noxious, rotten-egg-smelling gas, hydrogen sulfide.

While earth's creatures, in general, cannot survive in a hydrogen sulfide atmosphere, a few species of bacteria have evolved to thrive on it. Eschewing sunlight, these minute organisms derive energy to grow and reproduce by chemically oxidizing hydrogen sulfide through a process called *chemosynthesis*. Not surprisingly, thermal-vent waters harbor teeming clouds of such bacteria, whose extensive organic production parallels that of the minute, floating green plants in surface waters, and which provide the basis for the thermal-vent food chain (Figure 10.18).

FIGURE 10.17. Scenes and life forms associated with thermal vents of the Galapagos Ridge. (a) A black "smoker." (Photo by Dudley Foster.) (b) Vestimentiferan worms. (Photo by Jack Donnelly.) (c) Brachyuran crabs. (Photo by Robert Hessler.) (d) Fish and sampling apparatus. (Photo by Ken Smith.) (All photographs courtesy of the Woods Hole Oceanographic Institution.)

(a)

(b)

(c)

(d)

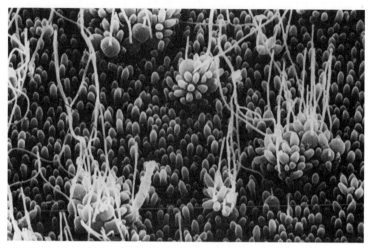

FIGURE 10.18. Sulfur bacteria from the Galapagos Rift thermal-vent system. (Courtesy of Woods Hole Oceanographic Institution.)

The giant clams apparently ingest these bacteria directly, filtering them from sea water which passes over their gills. Fish and crabs nibble and scavenge living and dead tissue in the vent community. In contrast, some of the vent worms (Figures 10.17 and 10.19) have opted for a more complex nutritional solution, obtaining their food from a mass of bacteria permanently housed in their body cavities. The bacteria produce chemosynthetically; the worms absorb the excess nutrients. By one analogy, an equivalent mutualistic relationship in human beings would require our housing a garden of green plants in our stomachs into which sunlight streams through a transparent window.

Chemosynthesis occurs in terrestrial and shallow-water environments as well but is less significant to the economy of these ecosystems. Exploration has shown that thriving thermal-vent communities may be fairly widespread on the sea floor. If so, we may have to significantly revise our estimates of the nature and extent of productivity in the deep sea.

Thermal-vent communities continue to pose a number of mysteries: Since the vents themselves are ephemeral, and because such complex, vent-specific biological adaptations probably take eons to evolve, how do vent residents find and move to new vents when their old ones expire? How do macroscopic vent creatures tolerate noxious hydrogen sulfide? Why have other food-limited deep-sea species not joined well-nourished vent communities? What peculiar metabolic adaptations allow vent organisms to survive in their unique environment?

To help solve these mysteries and others, *Alvin's* scientists carefully retrieve intact samples of vent life with the submersible's mechanical arms, scoops, tows, and a vacuumlike "slurp gun." Baited fish and crab traps (Figure 10.17) are placed on the sea floor

FIGURE 10.19. Organisms from the thermal vent system of the Galapagos Ridge. (a) Large clams. (b) Vestimentiferan worm. (Courtesy of Woods Hole Oceanographic Institution. Photographs by Jack Donnelly.)

(a)

(b)

to capture the more motile species. These will be recovered later. On return to the surface, collected organisms will be probed, measured, dissected, and analyzed by scientists from a number of research institutions. The work begins immediately in the laboratories of research vessels accompanying *Alvin* to the dive site.

Only a decade ago, what scientist could have imagined the spectacle of deep-water thermal-vent communities? Without doubt, myriad additional discoveries await our exploration of the depths. The deep sea is slowly yielding its secrets, but there is much more to learn. If the exciting discoveries of recent years are any indication of the future, a vital and fascinating odyssey lies ahead.

SUGGESTED READINGS

BALLARD, R. D. 1975. Dive into the great rift. *National Geographic* 147:604-15.

BRUNN, A. 1957. Deep-sea and abyssal depths. Chapter 22 (pp. 641–672). In Hedgpeth, J. E., ed. *Treatise on Marine Ecology and Paleoecology*, vol. I. Ecology. Memoir 67. The Geological Society of America. Boulder, Colorado: The Geological Society of America, Inc.

DIETZ, R. S. 1962. The sea's deep scattering layers. *Scientific American* 207:44-50.

EDMOND, J. M. 1982. Ocean hot springs: a status report. *Oceanus* 25:22-28.

EISELEY, L. 1957. The immense journey. New York: Vintage.

Galapagos Biology Expedition Participants. 1979. Galapagos '79: initial findings of a deep-sea biological quest. *Oceanus* 22:2-10.

GRASSLE, S. F. 1982. The biology of hydrothermal vents: a short summary of recent findings. *Marine Technology Society Journal* 16:33-38.

HARDY, A. 1965. *The open sea: its natural history*. Part I: *The world of plankton*. Chapter 12. Boston: Houghton Mifflin.

HEEZEN, B. C., AND HOLLISTER, C. D. 1971. *The face of the deep*. New York and London: Oxford University Press.

IDYLL, C. P. 1976. *Abyss*. New York: Crowell.

MACLEISH, W. H. 1978. The deep sea. *Oceanus* 21(1). (A collection of articles.)

MARSHALL, N. B. 1954. *Aspects of deep-sea biology*. New York: Philosophical Library.

MARSHALL, N. B. 1980. *Deep sea biology: developments and perspectives*. New York: Garland SPTM Press.

ROWE, G. T., ed. 1983. *Deep-sea biology*. New York: Wiley-Interscience.

WOODS HOLE OCEANOGRAPHIC INSTITUTION. 1982. Hot vent life forms and sea-floor geology. *Oceanus* 25:28-29 (A photographic essay.)

Chapter Eleven

Life in the Sea III:
The Coastal Zone

DO YOU REMEMBER YOUR FIRST INTRODUCTION TO THE OCEAN environment? Undoubtedly, it took place at the seashore, perhaps at a bustling harbor or a crowded summer beach. What were your impressions? Did you, like Keats's explorer, stare upon the scene "with a wild surmise," or was the magic of the moment marred by the signs and sounds of human commerce?

A visit to the coastal zone can, in fact, be an ambivalent experience. The narrow fringe separating land from sea is a region of contrasts. Here a quiet scene of incomparable beauty is splashed upon the landscape like a vibrant, surreal painting. There toxic wastes, of a civilization long used to casual dumping, seep into a bay already too fouled to support a fraction of the life it once cradled. What soul can fail to be stirred by the verdant mystery of a salt marsh in the early dawn mists? Can any of us not be outraged by the bloated bellies of rafts of dead fish, no longer flashing in quicksilver schools?

The waters of the coastal zone constitute but a fraction of the oceans, but they have an importance far out of proportion to their volume. They can be highly productive, rivaling the most efficient agricultural systems in annual food output. They are important nursery areas for juvenile fish, many of which spend their adult lives at sea. They buffer the landscape against the ravages of an angry sea. They provide humans with prime recreational sites and sheltered harbors for commerce and industry. They are worth spending some time with.

COASTAL FEATURES AND TERMINOLOGY

Before discussing specific coastal environments it is a good idea to have some familiarity with accepted terminology for the major features and landforms characterizing the boundary between sea and shore. These features are depicted in Figure 11.1 and are defined as follows:

> *shore:* region between low-tide level and the highest point affected by storm waves.
> *coast:* landward limit of shore inland to recognizable marine features like upper portions of estuaries and former shoreline levels.
> *coastline:* relatively fixed boundary between the shore and the coast.
> *shoreline:* boundary between land and water (changes with the state of the tide).

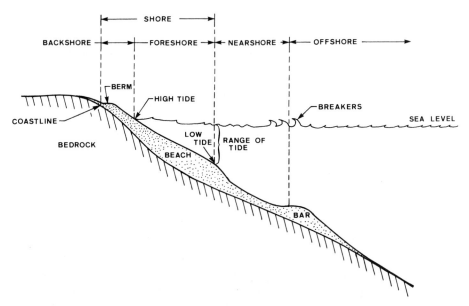

FIGURE 11.1. Coastal terminology.

backshore: region between high-tide level and the coastline.
foreshore: region between high- and low-tide levels.
nearshore: region between low-tide level and breaker zone.
offshore: from breaker zone seaward.
range of tide: vertical distance between mean high tide and mean low tide.

CHARACTERISTICS
OF COASTAL ENVIRONMENTS

Major Types and Shaping Forces

There is a considerable variety of coastal environments, each with its own spectrum of conditions and unique problems for resident and visiting organisms. Six specific coastal environments will be discussed in some detail in this section. These are *estuaries, coastal wetlands, beaches, rocky intertidal zones, subtidal benthic environments,* and *coral reefs.* Let's have a brief look at the physical forces which shape these environments and influence the structure and topography of the coastal zone in general.

The coastal landscape is molded and maintained by the deposition and erosion of sediments and by tectonic activities. In regions where sediments accumulate faster than they wash away, shoreline features build up. Examples are sand bars, barrier beaches, and deltas. Where sediment removal exceeds deposition, severe erosion can dramatically alter the face of the coast. Tectonic disturbances which affect coastal topography include volcanic activity, earthquake-caused slumping, and vertical movements of the continental crust.

Effects of Sea-Level Changes

In the long run the most important factor influencing the coastal landscape is sea level. In areas where sea level remains relatively constant over long periods of time, the coast is less subject to alteration than where sea level fluctuates. Where sea level rises gradually, the major characteristics of a shoreline generally remain largely intact, though the entire shoreline retreats gradually inland. Rapid rises in sea level inundate coastal features and may result in severe erosion. A lowering of sea level, while not necessarily physically destructive, can have an equally dramatic effect on the landscape by exposing formerly submerged areas of the seabed.

Sea-level changes can result from short-term phenomena or long-term processes. Variations occurring on a time scale of hours are attributable to tides, storms, and in semienclosed shallow embayments, excessive freshwater runoff from land. Longer-term variations in sea level, usually occurring on a time scale of millennia, may be of *eustatic* origin (i.e., resulting from changes in ocean volume) or attributable to major tectonic activities.

Eustatic phenomena which result in rising sea level include the melting of glacial ice (formerly "locked up" on continents), general warming of the oceans (causing expanded water volume), and the addition of "juvenile" water, newly issued from the earth's mantle (see Chapter 5). Eustatic causes of lowered sea level include increased glaciation (when evaporated sea water accumulates on continents as snow and ice) and cooling of the oceans, which results in contraction of their volume.

Tectonic activities which result in long-term sea-level change include continental drift (which repositions land masses on the earth's surface and changes the shape of ocean basins); geological uplifting or slumping of coastal land masses; and the *isostatic adjustment* of continents in response to the removal or addition of a large weight of glacial ice. *Isostasy* is the principle whereby portions of the earth's crust, including the continents, "float" in equilibrium or balanced relationship to each other in the denser, plastic mantle below. Like an iceberg, the deeper the crust extends into the mantle, the higher the uppermost elevation of the crust will reach. Conversely, like an iceberg, the more weight added (e.g., by ice and snow) to the elevated portion of the crust, the deeper the crust will sink into the mantle.

The isostatic response to glaciation and glacial melting is a longer-term process than the eustatic response to the same phenomena. For example, over the short term, increased glaciation will lower sea level because water is temporarily removed from the oceans but, over the long term, sea level may actually rise because the increased weight of snow and ice will push a continent deeper into the mantle.

General Characteristics
of Coastal Zone Environments

In marked contrast to the relatively stable conditions of deep offshore waters, the environment of the coastal zone is subject to enormous variability. This is primarily attributable to relatively shallow water depth, the influence of the terrestrial land mass, and the strong influence of the rising and falling tides.

Temperature fluctuations occur on both a daily and seasonal basis. Salinity levels are equally volatile. Heavy rain or excessive river runoff in shallow coastal waters can drastically dilute salt concentrations in a matter of hours, while a prolonged period of drought, when evaporation greatly exceeds precipitation, can turn a shallow, semilandlocked water body into a *hypersaline* lagoon in the course of a summer.

Other chemical properties may also undergo significant short-term variations. Concentrations of dissolved gases, including life-sustaining oxygen, fluctuate with water temperature, turbulence, tidal exchange, and biological activity. The pH levels can also undergo major changes, particularly in response to significant changes in CO_2 content (see Chapter 5). Nutrients often undergo fluctuations from undetectable levels to excessive concentrations, which may stimulate nuisance growth of algae.

The variations in physical and chemical properties experienced by shallow, coastal waters are not only dramatic, they are also largely unpredictable. In consequence, organisms which inhabit these rigorous environments have to be well adapted to a wide range of conditions.

With this background we are ready for a tour of coastal environments. We'll review their origins, physical characteristics, and environmental conditions and have a look at the forms, adaptations, and ecological interactions of their dominant organisms.

MAJOR TYPES OF COASTAL ENVIRONMENTS

Estuaries

Our first stop is an estuary (Figure 11.2). While most of us are familiar with the term *estuary*, we would all be hard pressed to define it, even though perhaps more has been written about estuaries than any other marine environment. Of all oceanic regions they have, arguably, the most direct importance for humans. Their relatively shallow depths, usually ample nutrients and, in most cases, thorough mixing make them highly fertile and productive regions. They afford protection and adequate nutrition for both juvenile fish and invertebrates and are, consequently, important nursery areas. It is estimated that three fourths of all commercial fish species in the

FIGURE 11.2. Slocums River estuary, South Dartmouth, Massachusetts. (Photograph by James Sears.)

United States depend on estuaries for all or part of their life history. Their economic value stems from the fact that estuaries are frequently good harbors and, therefore, important population centers. One third of the U.S. population lives and works on or near estuaries. Seven of the world's ten largest cities border estuaries.

Trying to generalize, let us think of estuaries as partially landlocked bodies of water in which fresh water and sea water converge and mix with measurable dilution of sea-water salinity. By this description an estuary could be represented both by a tiny stream-fed coastal pond with a narrow opening to the sea and by a vast embayment linked to major rivers (such as the Chesapeake Bay estuarine system). Clearly, estuaries are as diverse as they are similar: no two are exactly alike. They differ with respect to their origins, their geomorphology, and their patterns of circulation. These differences are instructive, so let's have a closer look at them.

CLASSIFICATION OF ESTUARIES. Historically, researchers have attempted to identify and categorize specific types of estuaries in terms of their special characteristics. One classification scheme is based on origin. Four basic estuary types are recognized.

A *coastal plain estuary* results from the inundation of a river mouth by the invading sea. Consequently, these estuaries are often referred to as *drowned river valleys*. Typically, the volume of the *tidal prism* (the amount of sea water exchanged in the estuary during a tidal cycle) is much greater than the fresh water influx. Coastal plain estuaries are exemplified by Chesapeake Bay.

Other estuaries are formed where former glacial ice deposits lay on the seaward end of river valleys. The pressure resulting from the weight of ice depressed the river bed, deepening the river valley in that region. When the ice melted, the rising sea filled in, mixing with river water. These estuaries are called *fjords*. They are common in mountainous northern regions subject to past glaciation (e.g., Scandinavia) and are characterized by having steep rocky sides, deep embayments, and a shallow sill at the mouth representing the nondepressed seaward extent of the former ice sheet. The sill restricts circulation between the ocean and the estuary so that the deep waters of the semienclosed water body frequently stagnate near the bottom.

Bar-built estuaries are also drowned river valleys but are distinguished from coastal plain estuaries by being subjected to considerable sedimentation. Characteristically, currents which travel parallel to the shore deposit sandy sediments in the quiet waters at the entrance of an estuarine embayment. Eventually, a sand bar grows partially across the mouth of the bay, modifying the estuary in the process. Such estuaries are often referred to as *lagoons.*

Occasionally, estuaries may result from disturbances in the earth's crust which create a barrier between coastal and offshore waters. These *tectonic estuaries* can form, for example, when earthquake-caused slumping or a volcanic eruption in the coastal seabed creates a sediment trap which eventually acts as a barrier to part of the coastline.

CIRCULATION PATTERNS IN ESTUARIES. Water circulation is perhaps the most important characteristic of estuaries, as it has major effects on the physical, chemical, geological, and biological properties of these coastal water bodies.

Estuarine circulation patterns, or *hydrodynamics*, are a function of the relative amounts of fresh water and tidal flow, bottom topography, physical dimensions, and short-term meteorological events like wind and storms. Circulation patterns are primarily influenced by the degree of mixing between fresh and salt water. This in turn determines the distributions of substances and organisms in estuaries.

Salinity is the property which most readily reflects the mixing characteristics of an estuary. All estuaries have a general gradient of increasing salinity from their *head* (which is the source of fresh water) to their *mouth* (which represents the interface with the fully marine environment). However, the vertical distribution of dissolved solids varies considerably among (and within) estuaries.

In *well-mixed estuaries*, a large tidal exchange ensures thorough mixing between sea water and fresh water. Sampling would reveal a nearly uniform vertical distribution of salinity at any point in the estuary but gradually declining salinity from estuary mouth to head (Figure 11.3c). Mixing is often facilitated by shallow depths. Well-mixed estuaries (sometimes referred to as *Type I estuaries*) are exemplified by Delaware and Chesapeake Bays.

Partially mixed estuaries (or *Type II estuaries*) experience less tidal exchange relative to the volume of incoming fresh water. Because of its lower density, the land-derived fresh water flows seaward over the top of the intruding ocean water (Figure 11.3b). Still,

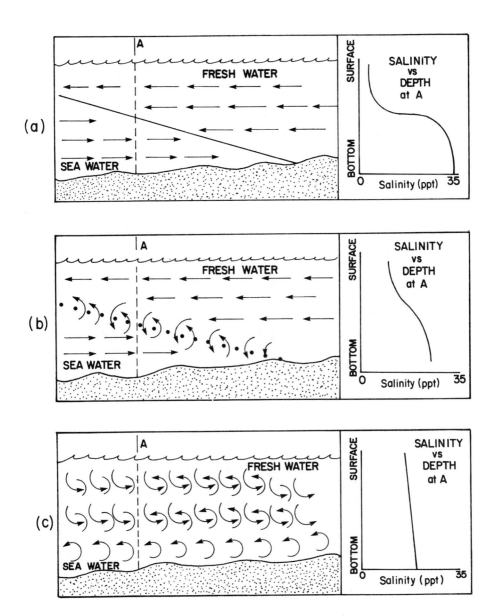

FIGURE 11.3. Estuarine circulation patterns. (a) Salt wedge estuary. (b) Partially mixed estuary. (c) Well-mixed estuary. (From Weiss, H. M., and Dorsey, M. W. 1979. *Investigating the marine environment*. Groton, Conn.: Project Oceanology.)

tidal influx is sufficiently strong to cause significant mixing of sea water into the overlying fresh water at the boundary between the two water masses. This process results in gradually increasing salinity in the surface waters flowing out. San Francisco Bay and Puget Sound are partially mixed estuaries.

In some deep estuaries, where exchange of ocean waters is impeded by a barrier, an intact sheet of fresh water flows seaward at the surface with little or no mixing with sea water below. These estuaries are referred to as *stratified* (or *Type III*) estuaries because of the vertical layering effect of water masses of different salinities. They are typified by fjords.

In other estuaries poor mixing results, not because of a seaward barrier, but due to a large volume of river flow relative to tidal influx. Such estuaries are characteristically shallow, allowing fresh water to dominate the circulation. A tongue of sea water creeps in along the bottom and undergoes slight entrainment into the voluminous fresh water flow above it (Figure 11.3a). Consequently, such water bodies are referred to as *salt wedge estuaries*. Often, the large river flow transports considerable quantities of sediments from the continent to the coastal zone seaward of the river mouth, building up *deltas* in the process. The Mississippi River estuary is a good example of a salt wedge estuary.

The degree of mixing in an estuary has important consequences for introduced substances like sediments, nutrients, and pollutants. In poorly mixed estuaries such substances tend to settle to the bottom and accumulate within the tidal prism. There they are retained within the estuary, moving back and forth along the bottom as the tides come and go. For this reason, many estuaries act as *sediment traps* and *nutrient traps,* and their productivity may be correspondingly stimulated. Unfortunately, for the same reasons, estuaries may also act as pollutant traps.

PROBLEMS AND ADAPTATIONS OF ESTUARINE ORGANISMS. Life is not easy in an estuary. A motile organism will encounter a great range of salinities as it travels the length of the water body. Similar variations may be experienced by plankton as they move vertically in the water column. Sedentary organisms like seaweeds, clams, and mussels are even more vulnerable to fluctuating salinity, experiencing a range of conditions from near-fresh water to full-strength sea water within a single tidal cycle. Pity the attached benthic organism caught in a downpour at low tide or enduring extraordinarily high salt concentration in a rapidly evaporating shallow pool on a searingly hot midsummer day.

Variations in temperature may be equally dramatic. During the summer, sea water is usually significantly colder than incoming river water; in winter, the pattern is reversed. In all seasons, an important consequence of this temperature difference is the thermal fluctuation occurring during a tidal cycle. Further, the shallow water of estuaries responds much more quickly than deeper ocean water to atmospheric temperature changes. Thus, large ranges in air temperature, particularly characteristic of temperate latitudes during spring and fall, will be reflected in major diurnal temperature changes in estuarine waters. The shallower the water, the greater the changes; exposed organisms will undergo the most severe variations of all. The seasonal temperature changes experienced by estuarine organisms may be even more profound. Who among us, for example, has not marvelled at the resiliency of an ice-choked embayment, locked in the grip of a winter freeze?

While fluctuating temperature and salinity are perhaps the most important stresses experienced by estuarine organisms, they are by no means the only ones. The considerable quantities of sediments carried into these water bodies result in high turbidity, with attendant reductions in light intensity, and frequently excessive siltation of the benthic environment. Siltation is a particularly acute problem for living organisms, as it fouls and clogs respiratory surfaces and feeding devices (particularly of filter feeders), smothers sedentary benthic forms, and reduces the amount of adequate settling surfaces for the spores of algae and the larvae of epifauna.

Other problems commonly encountered by estuarine organisms include infrequent oxygen depletion in the water column (primarily a function of poor mixing or tidal exchange) and exposure to trapped pollutants. It should be noted that sediment oxygen depletion commonly occurs at the sediment-water interface in most benthic environments, even when the overlying surface waters are well mixed. Estuarine sediments are typically *anoxic* (lacking oxygen) only a few centimeters below the surface.

The rigorous and largely unpredictable conditions prevailing in estuaries mean that relatively few species inhabit these marine environments. However, those that have successfully adapted tend to do very well because of the protection and nutritional benefits af-

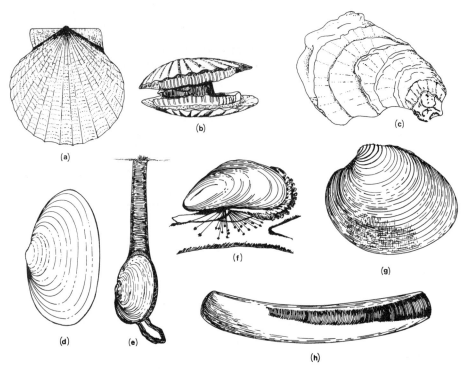

FIGURE 11.4. Estuarine bivalve mollusks. (a) and (b) Bay scallop, *Aequipecten irradians*. (c) Oyster, *Crassostrea virginica*. (d) Soft-shell clam *Mya arenaria*. (e) *Mya arenaria* with siphon extended. (f) Blue mussel *Mytilus edulis*. Note byssal threads attaching mussel to hard surface. (g) Hard-shell clam *Mercenaria mercenaria*. (h) Razor clam *Ensis directus*. (Redrawn from Arnold, A. 1901. *The sea-beach at ebb tide*. Reprinted by Dover, New York.)

forded by these semienclosed coastal habitats. As a consequence, estuaries are usually characterized by low species diversity but high productivity and abundance of resident organisms.

What adaptations enable estuarine organisms to survive—and even thrive—in their harsh environments? They are numerous and generally characterized as morphological, physiological, or behavioral. Let's examine some of the most important ones.

The impermeable shells of mollusks are perhaps the most obvious morphological adaptation aiding estuarine organisms. To a large extent, because of restricted or nonexistent motility, these organisms avoid physiological trauma simply by sealing themselves off in their shells. Common estuarine mollusks are depicted in Figure 11.4 and include the familiar and edible infaunal and epifaunal *bivalves* (possessing two shells hinged together on one side), represented by the hard-shell clam (*Mercenaria*), the soft-shell clam

(*Mya*), oysters (*Crassostrea* and *Ostrea*), and mussels (*Mytilus*), as well as gastropod mollusks like periwinkles (*Littorina*) and the rock-hugging limpets and slipper shells (*Crepidula*).

Additional important morphological adaptations include devices to minimize the potential for siltation-induced clogging of respiratory structures.

The most widespread adaptations allowing resident organisms to tolerate the extreme conditions of estuaries are physiological adaptations to fluctuating salinities. For a general review of responses of marine organisms to salinity variations, please refer to Chapter 9. A large proportion of estuarine organisms (with the notable exception of most bivalve mollusks) are osmoregulators. For example, higher crustaceans, including estuarine crabs, possess the well-developed ability to control the quantities of water, ions, and even free amino acids within their cells in response to external salinity.

Soft-bodied polychaete worms, including those of the genus *Nereis*, are both osmoconformers *and* osmoregulators. Their first response to changing salinity is to allow their internal fluids to match those of the outside environment. However, should external salinity drop too low, threatening rupture of internal cells because of excessive water absorption, the animal initiates osmoregulation.

For a number of estuarine organisms, the most reasonable response to the fluctuating conditions of their environment is simply to escape them. As drastic salinity or temperature changes occur in overlying surface waters, mud-dwelling infauna burrow deeper into the substrate: conditions in the *interstitial waters* of the sediments are comparatively stable. Other, more motile animals migrate up and down the estuary in response to ebbing and flowing tides. This behavior enables them to continuously occupy the same water mass with its comparatively constant conditions. Still other organisms have life histories which allow different developmental stages to take advantage of the most appropriate conditions for that stage. For example, euryhaline blue crabs (*Callinectes sapidus*) spend their adult lives in estuaries, but their larval forms, which are less tolerant of varying salinities, develop in the open ocean. On the other hand, the protected, nutritionally replete waters of estuaries provide nursery grounds for many juvenile fish whose adult stages live offshore.

ESTUARINE PRODUCTIVITY AND FEEDING ECOLOGY. In contrast to the open sea, whose food webs are based on grazing on planktonic primary producers, estuaries provide a great variety of food sources

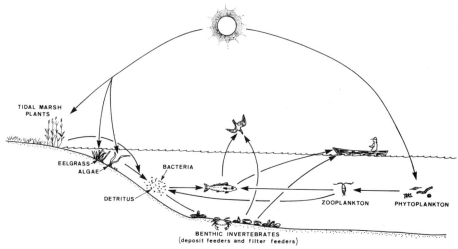

FIGURE 11.5. Estuarine-tidal marsh food web.

for consumer organisms. These include phytoplankton, attached macroscopic aquatic plants, and other organic matter carried in by tides from surrounding marine environments or by rivers from terrestrial sources.

Detritus (the decomposing remains of vegetation, nutritionally enriched by the bacteria which break it down) also provides an important source of food for many estuarine organisms. A generalized estuarine food web is depicted in Figure 11.5.

Coastal Wetlands

Scattered around the periphery of the estuary like fragments of delicate yellow-green ribbon are narrow regions of vegetation which separate the marine and terrestrial environments. Outside of the tropics, these intertidal transitional zones are composed of *salt marshes* (Figure 11.6). The tropical equivalent of a salt marsh is a *mangrove swamp*. Let's take a tour of a salt marsh first. Our vehicle will be a canoe which will afford maximum accessibility with minimal noise in the shallow, muddy environment. The time is early fall and the location a tidal inlet of Chesapeake Bay.

Silently, we glide into the entrance of a narrow creek which meanders through the marsh like an aimless roadway through a landscape of wheat fields. Like a wheat field, the marsh is a flat monotony of grassy vegetation. Two species dominate. Along the water's edge, at the creek entrance, a band of waist-high grasses thrusts out of the shallows. This is *salt marsh cord-grass (Spartina*

FIGURE 11.6. Salt marsh in Little River estuary, South Dartmouth, Massachusetts. (Photograph by James Sears.)

alterniflora; Figure 11.7). Like all grasses, these salt-tolerant forms (i.e., *halophytes*) are flowering plants (*angiosperms*); scattered through the regiments of upright shoots are flowering stalks, each of which, this time of year, is topped by a tassel of seeds. Shedding and subsequent germination of the seeds help to spread the plants, but most new annual growth is achieved through the production of new upright shoots from the perennially persisting underground root system—a process known as *vegetative tillering*. The cord-grass zone constitutes the *low marsh* environment. Truly intertidal, it is alternately inundated and exposed during every tidal cycle.

Quickly, our canoe passes through the fencelike barrier of cord-grass. Surprised by our intrusion, a small flock of black ducks explodes out of the creek ahead. We have entered a very different environment. Here the grasses are thin-stemmed and shorter and lie in a thick mat on the broad expanse of flat marsh above the cord-grass zone. They are frequently pressed into whorls, as if a large animal had bedded there for the night. This grass is *salt-marsh hay* (*Spartina patens*; Figure 11.8), the dominant vegetation of the *high marsh* zone which is only irregularly flooded by the highest tides of the month. The high marsh is usually the largest portion of the marsh, forming a broad, flat band of halophytic perennial plants between the mean high tide zone and the terrestrial shrubs and trees beyond. *Spartina patens* not only resembles hay; it also substituted for that bovine food source in colonial times. Early settlers allowed their cattle to graze on the high marsh and harvested it for fodder.

FIGURE 11.7. The salt marsh cordgrass, *Spartina alterniflora*, in flower. (Photograph by James Sears.)

At ground level, marsh vegetation and topography appear quite uniform. This appearance is somewhat misleading. An elevated vantage point or overflight would reveal a colorful mosaic of plant species, even on the high marsh. This variety results from surprisingly small changes in marsh elevation relative to water level. Height differences of only a few centimeters can result in very different flooding and drainage patterns, with correspondingly significant variability in soil moisture content, salinity, and other chemical properties. Plant species which are adapted to the specific set of conditions in a particular *microenvironment* will inhabit that limited area. A characteristic zonation of plant species in a salt marsh is depicted in Figure 11.9.

If you were to fly over most U.S. marshes you would also see another feature interrupting the apparent ground-level uniformity of these coastal wetlands. Slicing through the marsh in haphazard fashion, narrow, straight waterways carve up the grassy flats like pieces of key lime pie (Figure 11.6). These canallike features are *mosquito ditches*, dug by the Civilian Conservation Corps during the New Deal era of the 1930s, ostensibly to facilitate the flooding and drainage of stagnant, insect-breeding pools of water on the high marsh. They have not worked especially well, particularly where they are poorly maintained. Improper maintenance can actually exacerbate the very drainage problems the ditches were designed to resolve. When well maintained they can supplement the role of the

FIGURE 11.8. Salt marsh hay, *Spartina patens*. (Photograph by James Sears.)

network of natural creeks in renewing sea water and supplying nutrients to the infrequently flooded high-marsh environment.

Much has been written about the value of salt marshes, some of it apocryphal. What can we accurately say about the ecological and economic worth of these beautiful coastal regions? To begin with, they are among the most fertile and protected components of an estuary and provide the nursery grounds for many of the fish and invertebrate species which depend on the estuary for the juvenile stages of their life histories. They provide habitat for a considerable variety of organisms which are intimately interdependent. They are

FIGURE 11.9. Generalized zonation of a salt marsh.

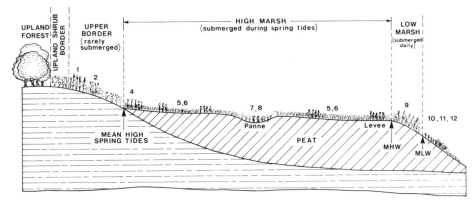

highly productive, producing as much organic matter in a year, without artificial fertilizer or energy subsidies from humans, as an equivalent area of lush farmland. They buffer valuable coastal property against the encroachment of the sea. They harbor large populations of game birds and sport fish to the lasting pleasure of hunters and anglers. Too often their worth is sacrificed through filling and subsequent development.

Marshes also act as important coastal sediment traps. As silt-laden tides sweep across these wetlands, emergent plants baffle the flow of water, causing suspended particles to settle out onto the marsh surface. With the accretion of sedimentary materials and gradually rising sea level, the marsh grows horizontally and vertically, its elevation finely tuned to the levels of the tides. Through much of their range, salt marshes do not grow in the winter, and the previous summer's growth does not completely decay before being buried by the remains of the next year's production. The partially decomposed remains of marsh grasses accumulate as a spongy *peat* which is largely held together by the extensive root systems of the living grasses.

While there is considerable growth of *Spartina* roots and rhizomes below the sediment surface, the annual growth of upright *Spartina* shoots accounts for 90 percent of a marsh's above-ground production of some 3 kilograms of dry organic material per acre per year. What is the fate of this nutritional windfall? Surprisingly, almost none of it is eaten directly: only a few insect species graze on the living stems and leaves of these grasses. The overwhelming proportion which survives the summer withers and dies in place at the end of the growing season. Water currents, storms, and ice then work their will, here breaking off small fragments, there tearing whole clumps from their marsh-surface moorings. Flotillas of plant litter are launched into ditches and creeks and sail off, in endless procession, into the bordering embayment. There, much of it accumulates in windrows at the water's edge.

Whether the dead plant material is carried off into surrounding waters or abandoned on the marsh surface, it is not left unattended for long. Decomposing bacteria quickly colonize the nitrogen-poor litter. Soon the dead plant material is converted into a nutritious form of detritus which is highly favored by a multitude of marsh scavengers, including shrimp, crabs, and snails (Figure 11.10). Thus, the grasses ultimately provide the basic energy source for the entire marsh system.

FIGURE 11.10. Salt marsh invertebrates. (Photograph by James Sears.)

The abundance and importance of marsh grasses tends to over-shadow the productivity of other primary producers in the marsh. However, microscopic algae, large seaweeds, and submerged grasses are often abundant in salt-marsh ecosystems. On your next trip to a salt marsh in the spring or early summer, look carefully at the base of the marsh plants, where you will likely find a dense mat of marsh algae.

There is considerable debate as to whether salt-marsh ecosystems (including bordering embayments) consume or retain all of the organic matter which they produce or whether significant quantities are "exported" into adjacent coastal waters. Indeed, the argument that marshes "fuel" the overall productivity of the coastal zone by pumping out vast quantities of decomposing grasses (through a process called *outwelling*) has been used as an important rationale for preserving these delicate systems. It is now apparent that such a generalization is probably too sweeping. The likelihood is that all marshes behave differently, some being exporters, some accumulating excess organic production, some internally consuming all that is produced, and some actually relying on imports of material to supply their energy needs.

FIGURE 11.11. A mangrove tree in the tropics. Note the prop roots. (Photograph by James Sears.)

In tropical areas, mangrove swamps replace salt marshes as the characteristic form of coastal wetlands. Though trees (Figure 11.11), not grasses, are the dominant vegetation, these systems are remarkably similar to salt marshes in their high productivity, intricate, detritus-based food webs, and sediment-trapping properties.

Beaches

Forsaking our canoe for our own two legs, we leave the salt marsh and head for an entirely different coastal environment—a beach (Figure 11.12). Most of us appreciate a beach for its recreational attractions, but have you ever stopped to consider the remarkable physical and biological properties of this environment? Let's kick off our shoes and have a closer look.

As we stand at the water's edge with the waves caressing our feet, it is hard to imagine a beach without sand. Could you conceive of one made entirely of old tin cans? In fact, such a beach exists on the south coast of England. The source of the cans is a large offshore dump; waves have transported the discarded containers to the coastal zone.

While admittedly unique, a tin-can beach conforms to the general definition of a beach as a strip of wave-worked materials lying between the breaker zone and the coastline. The variety of materials which makes up the beaches of the world includes the black "sands" of ground-up lava, the remains of shells and skeletons of marine organisms (the principal component of the gleaming white sands of

FIGURE 11.12. A beach in the tropics. (Photograph by
H. S. Parker.)

tropical isles), large cobbles and boulders, and the familiar quartz
and feldspar sediments of temperate latitudes.

As you scan the length of the beach, you probably take for
granted its characteristic, gradually sloping profile. Supposing the
months suddenly raced by, and you were standing here on a frigid,
spray-lashed, midwinter day. The beach slope would be steeper,
and the fine sands of summer would be replaced by coarser grains
and pebbles with an upper boundary of large cobbles.

Focus on the landward boundary of the beach for a moment.
The backshore region is characterized by the *berm*, a low, flat expanse which is the highest part of the beach and which gradually
merges with the steeper slope of the beach face, or *foreshore* (Figure 11.13). Again, a winter profile would be much different, with

FIGURE 11.13. Generalized profile of a beach. The solid line represents a
summer beach profile and the dashed line a winter beach profile.

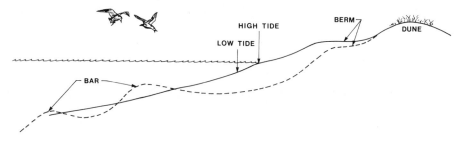

FIGURE 11.14. Sand dunes. (Photograph by James Sears.)

the berm considerably narrower and a sharper transition (the *berm crest*) between the berm and beach face.

What accounts for the differences between winter and summer beaches? High-energy winter-storm waves excavate the berm and erode fine sand particles from the beach face, depositing them in bars outside the breaker zone. The coarser-grained sediments are left behind. The gentler waves of summer redeposit this sand on shore, building up the smaller-grained sediments and increasing the slope of the beach face in the process. The seasonal processes of beach alteration are depicted in Figure 11.13.

Let's momentarily wander inland from the beach face. On many beaches you will immediately encounter a series of gentle sand hills overgrown with a variety of grass and shrub vegetation. These are the *dunes* (Figure 11.14); beaches which possess them are known as *barrier beaches* (Figure 11.15) because the dunes are a first line of defense for the shoreline against the ravages of the sea. Characteristically, a salt marsh and its estuary or a coastal lagoon lie on the protected, landward side of the dunes. A generalized profile of a barrier beach is depicted in Figure 11.16.

FIGURE 11.15. A barrier beach enclosing a coastal salt pond in South Dartmouth, Massachusetts. (Photograph by James Sears).

Barrier beaches are built where water-borne sediments are transported and settle out in shallow coastal areas. As the sediments accumulate they form *sand spits* which may eventually rise above the sea surface, allowing colonization by terrestrial vegetation. The importance of this vegetation cannot be overemphasized, for it stabilizes the entire barrier-beach system. Let's have a closer look at this role.

Barrier beaches, like beaches in general, are notoriously unstable environments. Even a brief visit to these areas on a relatively calm day will help you appreciate this. Surging waves and gusting winds constantly keep the fine sand grains in motion. Imagine the effect of storm winds and waves on this unconsolidated sediment. Without a baffling agent, the beach would soon be leveled, its sand transported into the lagoon behind.

Snow fences and dead Christmas trees make good baffling agents, but living vegetation does an even better job, sufficiently diminishing wind strength and water currents to allow sand grains to settle out. Dune grasses, including the common northeastern U.S. beach grass *Ammophila* (Figures 11.17 and 11.18), wild rye

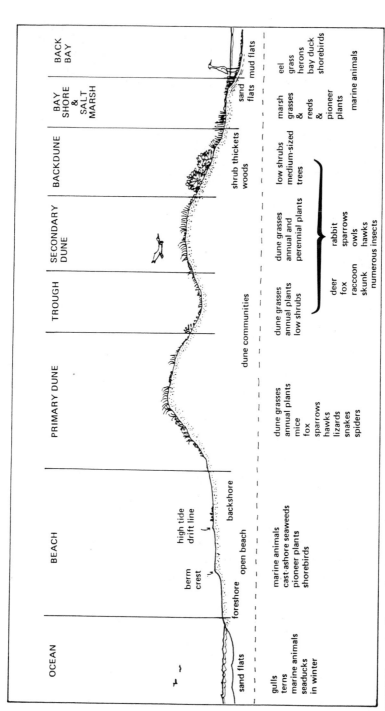

FIGURE 11.16. Generalized profile of a barrier beach. (From Fox, W. T. 1983. *At the sea's edge.* Englewood Cliffs, N.J.: Prentice-Hall. Drawing by Clare Walker Leslie. Reproduced with permission.)

FIGURE 11.17. The beach grass, *Ammophila*.
(Photograph by James Sears.)

(*Elymus*) in higher latitudes, and sea oats (*Uniola paniculata*) in the southeastern United States do a particularly good job of this. Watch a clump of beach grass carefully as the wind sweeps across the dunes. The upright stems trap the moving sand, which builds up into little hillocks at the base of the plants.

How can seemingly delicate grasses remain in place in such a dynamic environment? Gently scrape away the sand around the vegetation. If you were to dig down far enough you would find an extensive and deep network of roots and *rhizomes* (rootlike underground plant stems which can give rise to new roots and shoots) which anchor the plants in their unstable habitat and draw scarce moisture from fresh-water lenses deep within the dunes. Nutrients are also scarce in the beach sediments, which contain relatively little organic matter. To offset this, several species of dune vegetation are able to use atmospheric nitrogen by "fixing" it within their root systems (see Chapter 9).

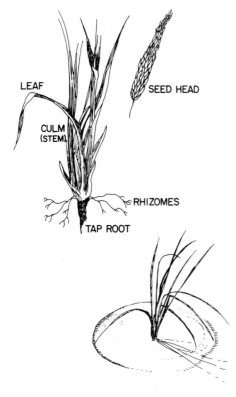

LEAF

SEED HEAD

CULM
(STEM)

RHIZOMES

TAP ROOT

FIGURE 11.18. Detail of American beachgrass, *Ammophila breviligulata*. (From Fox, W. T. 1983. *At the sea's edge.* Englewood Cliffs, N.J.: Prentice-Hall. Drawing by Clare Walker Leslie. Reproduced with permission.)

A healthy vegetation cover on a barrier beach ensures its continuous maintenance, even in the face of slowly rising sea level. The seaward portion of the beach may grudgingly give ground to the reclaiming ocean, but the entire beach-estuary system remains intact and gradually retreats inland or "rolls over." This natural process is accelerated during storms, when waves frequently break over the entire beach (a process referred to as *overwash*). However, only if sea level rises too fast or if humans interfere with the natural self-preservation capacity of barrier beaches do these coastal ecosystems and their associated salt marshes and estuaries become destroyed. Unfortunately, this interfering human hand is often felt.

Visit almost any beach in the coastal United States, and you will see the damage wrought by humans. Development is the most obvious villain. Widespread construction on barrier beaches destroys vegetation and associated wells deplete scarce supplies of fresh water. Denuded portions of the dunes can no longer stabilize moving sand. Wind erosion is more dramatic and overwashes more frequent. Eventually semipermanent channels may be cut through to the estuary, further accelerating the process of erosion.

The threat to barrier beaches is not limited to extensive development. Even recreational beaches are vulnerable. Parking lots, footpaths, and boardwalks cut a lethal swath through vegetation, making the affected area vulnerable to breaching. Even human attempts to stabilize or arrest natural beach migration, exemplified by sea walls, breakwaters, and groins, may accelerate the process of erosion.

Barrier beaches on the U.S. east coast have undergone unusually intense development. Ominously, most of the construction has occurred during the past 20 years, a period which has coincided with an unusual lull in storm activity. The lessons of the killer hurricanes of 1938 and 1954, when many beaches in the Northeast were literally swept clean of houses, appear to have been forgotten.

In contrast to human beings, who are not particularly suited to a beach existence (surfers and beach bums notwithstanding), a number of organisms are clearly at home in this unstable environment. This may seem surprising at first. Life is not easy on a beach. For one thing, imagine trying to live in an environment which is constantly in motion. Every gust of wind and passing wave churns up the substrate, suspends it in the water, and redeposits it elsewhere.

Consider also the problem of moisture on a beach. Aquatic organisms inhabiting the sand may face the possibility of drying out. Water retention by beach sediments is largely a function of particle size. In coarse-grained beaches, wave-delivered water may percolate so rapidly through the sand grains that resident organisms are vulnerable to desiccation. Porous sands also prevent the accumulation of organic matter, which is so important to the nutrition of organisms in a muddy bottom. Burrowing (the preeminent lifestyle of sandy shore organisms) is also more difficult in coarse sediments. Finally, the intermittent submersion and draining of beach sands results in rapidly fluctuating physical and chemical conditions in the top 10 to 15 centimeters of the *interstitial* (between grains) environment. Below that depth lies a relatively permanent and environmentally stable layer of insulated sea water, which, because of its greater density, is relatively resistant to mixing with fresh water, even during a downpour.

These rigorous conditions restrict the numbers of organisms which inhabit the foreshore of beaches. A glance around the beach

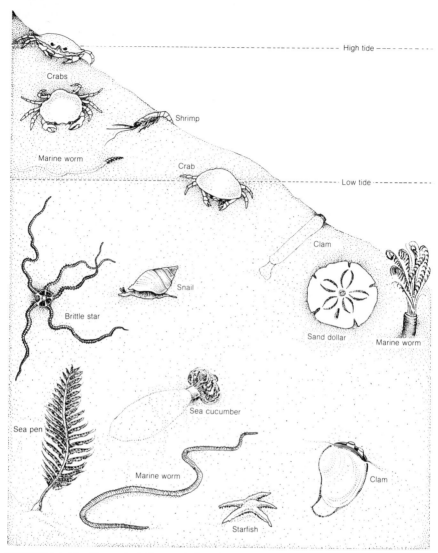

FIGURE 11.19. Characteristic sandy beach zonation and organisms. (From Duxbury, A. C., and Duxbury, A. 1984. *An Introduction to the world's oceans.* Reading, Mass.: Addison-Wesley.)

will confirm this. While a variety of uprooted seaweeds have been tossed up by the waves, large marine plants do not live on the beach itself because they cannot get a firm grip in the shifting sands. Similarly, most large benthic invertebrates, such as mussels, barnacles, and starfish, so common in rocky intertidal areas, are absent here, again because of the lack of a firm bottom.

Nonetheless, a careful look below the surface of the sand reveals a surprising variety of organisms which are well adapted to the difficult conditions of their environment. Mostly, these live underground. They include relatively large animals like clams and crabs; smaller but still visible organisms like amphipods, isopods, and a variety of worms; and microscopic creatures, including benthic diatoms, protozoans, bacteria, and fungi. Some of the more common species are depicted in Figure 11.19.

How do beach organisms survive in their unstable environment? Let's look at some of the more notable adaptations. Look closely at a tiny depression in the sand which appears to have a bit of flotsam caught in it. In fact, the "flotsam" turns out to be the eyes and feathery antennae of a mole crab (*Emerita*). This animal has actually capitalized on wave energy by burrowing into the sand tail first and filtering microscopic food particles from receding waves with its protruding antennae. Nearby, another invertebrate, the bivalve *Donax*, has adopted an even more remarkable means of using wave energy to its benefit. In essence, it maintains its position relative to tide level by "deliberately" exposing itself to breaking waves. During incoming tides, it emerges from its burrow just before a wave breaks, thus getting a free ride up the beach. It then burrows back into the sand before the wave recedes and waits for the next wave. During outgoing tides, it emerges from the sand just after a wave breaks, enabling the retreating wave to carry it back *down* the beach.

Burrows are a characteristic feature of the largest beach-dwelling organisms and facilitate the obtaining of food and oxygen as well as providing a reasonably protected habitat. A number of smaller beach organisms, exemplified by amphipods and isopods, do not live in burrows but live interstitially within the sand grains. These organisms characteristically feed by filtering minute particles from the interstitial water or by scraping microscopic algae and bacteria from the particles themselves.

Muddy intertidal areas have many of the same features as sandy beach environments. They are alternately exposed and flooded by the retreating and advancing tide, they have a characteristically shallow slope, they may experience wide variations in temperature and salinity, and their substratum is quite unstable. However, organisms living in mud flats must cope with sediment conditions which are siltier and frequently less oxygenated than those on sandy beaches. In general, they cope quite well. For example, the lugworm *Arenicola* lives in a U-shaped burrow (Figure 11.20). It feeds by ingesting sediment at one end, assimilating the

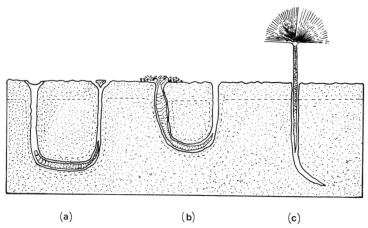

(a)   **(b)**   **(c)**

FIGURE 11.20. Burrowing habit of three muddy bottom animals. (a) *Arenicola marina.* (b) *Amphitrite johnstoni.* (c) *Sabella pavonina.* (From Meadows, P. S., and Campbell, J. I. 1978. *An introduction to marine science.* Glasgow: Blackie & Son, Ltd. Reproduced with permission.)

nutrients, and excreting the nondigested pellets at the other end. Clams extend their siphons out of the burrow entrances, enabling them to draw in food-laden, oxygenated water even though the rest of the animal may be located in an anaerobic environment some distance below the sediment surface (Figure 11.4). This strategy is employed by infaunal bivalves in all nonrocky benthic environments.

FIGURE 11.21. The rocky intertidal. (Photograph by James Sears.)

Let's leave the beach now and walk to a nearby rocky headland. Even from a distance the booming waves and wind-blown spume give notice that this is a very different environment from the relatively placid beach (Figure 11.21). Unlike beaches, where wave energy is dissipated over a broad, gently sloping expanse, rocky shores meet the seas head on. Seemingly, the coast is conqueror, vanquishing its antagonist in a shower of heaven-sent spray. A closer look dispels this conclusion. Tireless legions of waves roil against the stubborn shore, and the rocks show tortured signs of wear. Here a crevice cleaves the granite surface; there a sand-scoured hollow, smoothed by the abrasions of a billion waves, harbors a hamlet of periwinkles.

Just as the environment of a rocky shore contrasts with that of a beach, so do its resident organisms face different problems from those inhabiting sand. Their substrate may be firmer, but the physical force of moving water is greater. It is difficult to believe that any living organism could survive such repeated battering. Yet look around the rock. Not only is life in evidence, it thrives in teeming profusion. Barnacles, seaweeds, snails, and mussels vie for space on the crowded rock surface, leaving scarcely a bare area available for new colonization (Figure 11.22).

FIGURE 11.22. Competition for space in the rocky intertidal (barnacles and algae. (Photograph by James Sears.)

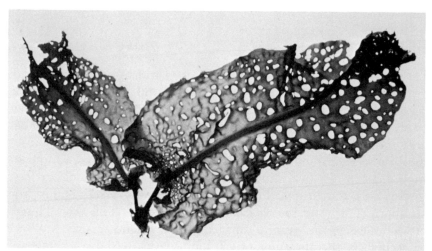

FIGURE 11.23. The kelp *Agarum*. The holes are natural to the plant. Note small holdfast at base of the two stem-like structures called stipes. (Photograph by James Sears.)

How do rocky-intertidal organisms even stay attached under such physically stressful conditions, let alone become so abundant? To some extent, nooks and crannies in the rock afford sanctuary, but that is only part of the answer. Examine any organism inhabiting this zone, and you will find that it is extraordinarily well attached to the rock. The rootlike *holdfasts* of seaweeds, including the familiar rockweeds and kelps (Figure 11.23), and the cementing accretions of barnacles are cases in point. Try pulling a mussel off the ledge. A mass of sticky, hairlike secretions called *byssal threads* (Figure 11.4) glue it firmly to the surface.

More motile animals, such as starfish, periwinkles, and sea urchins, are not such good clingers. They maintain their position by nestling in crevices and hollows, sometimes, as exemplified by *Strongylocentratus purpuratus*, a West Coast urchin, excavating their own depressions in the solid rock. Familiar organisms of the rocky intertidal zone are depicted in Figure 11.24.

Fortuitously, we have visited this rocky shore during an ebbing tide. As the waters recede a clear pattern of life forms is revealed on the exposed ledges: Several distinct horizontal bands of animals and plants are arrayed in vertical sequence from the upper intertidal to the water's edge (Figure 11.25). Each band has a unique assemblage of organisms which contrasts sharply with those above and below. What accounts for this characteristic *zonation* of the rocky intertidal area? The answer is revealed in the peculiar

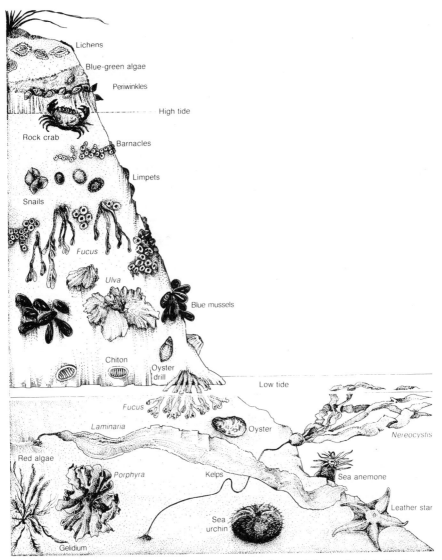

FIGURE 11.24. Characteristic zonation and organisms of the rocky intertidal. (From Duxbury, A. C., and Duxbury, A. 1984. *An introduction to the world's oceans.* Reading, Mass.: Addison-Wesley.)

conditions of this coastal environment and the adaptive responses of the resident species.

Battering by waves is not the only insult suffered by rocky-intertidal organisms. Because of their steep slope, rocky shores are subject to a greater range of tide than gradually sloping beaches. Consequently, a resident organism may spend a considerable portion of the day out of water. Imagine the extremes of temperature, salinity, and moisture endured by a plant or animal alternately sub-

FIGURE 11.25. Banded zonation pattern of the rocky intertidal. (Photograph by James Sears.)

merged and exposed under an infinite variety of weather conditions. Survival requires appropriate morphological, behavioral, and physiological adaptations. These include the impervious shells of mollusks and crustaceans, the mucilaginous, water-retaining, and temperature-buffering thick cell walls of rockweeds, the tendency of many intertidal organisms to inhabit moist, protected crevices or the subsurface layers of an insulating algal mat, and energy-conserving metabolic adjustments to the stressful conditions of exposure.

The uppermost limit of the intertidal zone, above the high-tide level, is the *littoral fringe*, or *splash zone*, so named because its exposure to sea water is limited to the spray from breaking waves. Splash-zone inhabitants include blue-green algae and lichens (which give the rock a characteristic black color) and a few snails. All of these organisms can endure long periods of desiccation. Characteristically, the high-water barnacle, *Chthamalus*, occupies a zone at the lower limit of the splash zone.

The intertidal zone, between Mean High Water of spring tides (MHWS; see Chapter 8) and Mean Low Water of spring tides (MLWS), is referred to as the *midlittoral zone*. It is dominated by the low-water barnacle (*Balanus*), mollusks (including *Mytilus*), and a few small rockweeds. In reasonably sheltered areas the rockweeds are generally the most abundant forms throughout the zone, but in more exposed areas strong wave action prevents their firm establishment.

The lower intertidal and upper subtidal region (the *sublittoral fringe*) is characterized by a broad band of kelps and, often, a considerable variety of red algae, whose special accessory pigments (which give them their distinctive color) make them well suited to the comparatively low light intensities here.

A generalized zonation pattern of rocky intertidal areas is depicted in Figure 11.24. The width of the various bands depicted in Figure 11.25 is a function of the range of the tide which, in some areas, may be only a few centimeters and in others (exemplified by the Bay of Fundy region in eastern Canada) up to 10 meters. Despite the width of the intertidal zone, the distinct patterns of zonation reflect, in part, the different degrees of tolerance of exposure by organisms in each horizontal band.

The stresses resulting from exposure and the different tolerances of resident organisms are not the only factors governing distribution in the rocky intertidal. A glance at this crowded seascape indicates that unoccupied rock surface is in short supply; there is considerable competition for living space. Organisms which, for one reason or another (e.g., superior tolerance to exposure, fast growth, unpalatability to predators, or high reproductive fecundity), have a competitive edge in a particular zone tend to overgrow the substrate there, crowding out other species. There is considerable interest and ongoing research in the ecology of rocky intertidal areas, but evidence to date suggests that physical factors (primarily exposure) govern distribution patterns in the upper intertidal, while biological factors (competition and predation) are principally responsible for determining who can successfully inhabit the lower intertidal.

Before leaving this fascinating coastal ecosystem, let's linger for a moment at a *tide pool*—a shallow pool of water trapped high above the receding tide (Figure 11.26). Crouched on our hands and

FIGURE 11.26. Tide pool. (Photograph by James Sears.)

knees, we carefully explore this microcosm of the ocean environment. Life is abundant here: A fringe of grass-green algae girdles the
border of the pool, and other seaweeds carpet its bottom. Periwinkles cluster in dark recesses below the water's surface. Ragged
swatches of barnacles thrust from the rock's surface like miniature
mountain ranges; from some, a feathery appendage juts from the
peak and sweeps the water for small food particles. Peer at the pool
more closely, your eyes just above its surface. Amphipods and isopods, crustacea resembling tiny shrimp, dart about, shadowed
closely by minnows which also make this pool their home. The
stalked eyes of a green crab protrude above the limp fronds of a seaweed. In sheltered crevices below these fronds, the dimpled arm of a
starfish and the sharp spines of a sea urchin are barely visible.

Tide pools offer a more stable environment than the alternately
exposed and covered surface of the intertidal rock. In that sense they
provide refuge for a variety of species which might not otherwise
survive high above low-tide level. Still, life is difficult enough in
these miniature habitats. Depending on the depth of the pool, the
duration of exposure to the elements, and weather conditions they
may undergo major changes in salinity, temperature, nutrient content, pH, and dissolved oxygen concentration before sea water is renewed with the next incoming tide.

Subtidal Benthic Environments of the Coastal Zone and Continental Shelf

Among the most productive marine communities are the subtidal
benthic environments of estuaries and the continental shelf. The invertebrate organisms of the shelf have been particularly well described. The marine ecologist Gunnar Thorson determined that
specific assemblages, or communities, of relatively sedentary infaunal and epifaunal bottom-dwellers are characteristic of specific
water depths and/or sediment types. For example, similar groups of
invertebrates (*parallel communities*) which are dominated by the
bivalve *Tellina* inhabit clean, sandy bottoms from the intertidal
zone to a depth of about 10 meters throughout the world. In contrast, communities dominated by the echinoderm *Amphiura* characteristically live in soft or silty sediments in deeper waters.

Benthic organisms inhabiting soft sediments often disturb the
bottom sediments through their burrowing or movement, leaving

conspicuous evidence of their activities. This *bioturbation* may benefit infaunal organisms by delivering oxygen and nutrients deeper into the sediments than would occur in its absence.

Benthic plants are extremely productive in some subtidal areas of the continental shelves or estuaries. For example, flourishing stands, or "forests," of *giant kelps* (large brown seaweeds of the genera *Macrocystis, Nereocystis,* and *Alaria*) dominate some coastal areas of the Pacific and South Africa and form the basis of a profitable harvesting industry (see Chapter 12). Some of these seaweeds reach lengths of 100 meters and grow at the rate of 25 centimeters per day. In the northwest Atlantic, the productivity of certain species of kelp of the genus *Laminaria* is among the highest for any plant on earth.

Though the seaweeds are confined to a narrow coastal fringe where light penetrates to the bottom, their contribution to oceanic productivity is disproportionate to the area they occupy because of their very high growth rates under favorable conditions. Though they inhabit less than 0.1 percent of the ocean's surface area, they may contribute up to 10 percent of the marine primary productivity. Kelp forests are also important because they provide habitats and shelter to a wide variety of organisms.

An equally productive group of benthic marine plants is the sea grasses. A number of genera exist worldwide, but two are particularly abundant and productive. These are *Zostera* (eelgrass) in temperate latitudes and *Thalassia* (turtlegrass) in tropical latitudes. Unlike the seaweeds and other algae, these submerged grasses are *higher plants,* possessing true leaves, stems, and roots, and reproducing through flowers and seeds. They are most abundant in shallow, protected estuaries and, like the kelps, are among the most productive kinds of plants on earth. Also like kelp forests, sea-grass communities provide shelter and habitat to numerous marine species. In addition, they play an important role in stabilizing estuarine sediments.

Coral Reefs

We will have to fly to our next coastal habitat, since coral reefs are restricted to the warm waters of the tropics or subtropics (Figure 11.27). Interestingly, individual corals may be found even in temperate-latitude waters, where winter water temperatures drop to a couple of degrees above freezing, but these corals never form reefs.

The reef-building corals of the low latitudes are most productive when water temperatures average 23 to 25° C (coral reefs never form below 18° C) and where ample sunlight penetrates the ocean's surface layer. The requirement for sunlight may seem surprising in view of the fact that the individual corals which make up the reef are animals—more specifically *coelenterates* (a phylum which is distinguished by the presence of stinging cells and which includes the jellyfish and sea anemones, to which they are closely related)—with a skeleton of calcium carbonate. However, if you were to carefully examine the tissues of these corals, you would find that most contain an abundant supply of minute, single-celled photosynthetic algae in their inner cells. These algae, called *zooxanthellae*, have a highly specialized and interesting relationship with the coral animal (the *polyp*) and are the key to the coral's productivity. Let's examine this relationship.

Corals which contain zooxanthellae are referred to as *hermatypic* (and those without the algae are known as *ahermatypic*). The algal-coral interaction constitutes a mutually beneficial *symbiosis:* The corals apparently provide nutrients and protection to the algae, and the zooxanthellae provide soluble organic molecules (photosynthetic products) to the tissues of the corals or may even be directly ingested by the animals. How helpful are the algae to the

FIGURE 11.27. Coral reef. Note several varieties of corals. (Photograph by James Sears.)

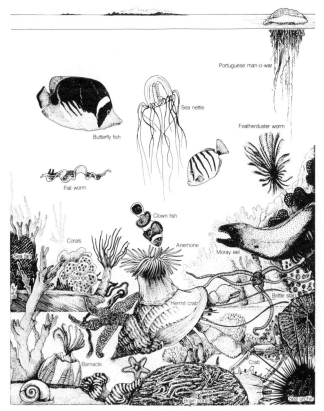

FIGURE 11.28. Characteristic organisms of a coral reef. (From Duxbury, A. C., and Duxbury, A. 1984. *An introduction to the world's oceans*. Reading, Mass.: Addison-Wesley.)

corals? The evidence is that they are extremely helpful, since the rate of *calcification* (production of the calcium carbonate skeleton) decreases with depth (and increases with light intensity), and since reefs are primarily constructed of hermatypic corals. In fact, a coral may live for months on algal byproducts if external food sources are not available.

Let's don skin-diving gear and snorkel over the surface of the reef at high tide. We'll also need a flashlight, since the best time to visit a reef is at night when its residents are most active. We'll also take along a geological pick and an open-mesh net bag for collecting samples.

Silently, we slip beneath the slight chop and glide over the reef. For the most part we leave the light off so as to disturb the reef organisms as little as possible. The intermittent illumination reveals a startlingly prolific community (Figure 11.28)—coral reefs are among the most productive and diverse ecosystems on earth.

The high productivity of a coral reef is somewhat surprising, since waters surrounding coral reefs are characteristically deficient in nutrients and produce only about 1 percent as much organic matter as the reefs in the course of a year.

The high diversity of coral reefs is due to the large number of niches created by the reef's complex physical and biological environment. More than 3000 different animal species may be found inhabiting some large reefs. Most of these have well-developed symbiotic relationships with other animals or plants on the reef. These relationships may be extraordinarily specialized, as exemplified by the coral-zooxanthellae *mutualism* (a symbiotic relationship between organisms in which both individuals benefit from the interaction). An even more astonishing symbiosis is displayed by the *coevolutionary* (the simultaneous development of adaptations in interacting populations) adaptations of "cleaner fish" and their hosts. The cleaners feed on external parasites of larger fish without danger of being eaten in the process. Infested fish actually line up at conspicuous "cleaning stations," where an exchange of mutually recognized signals allows the cleaners unobstructed access to the host fish—often even to the latter's gills and mouth where parasites may be particularly abundant. In a bizarre variation on this theme, some ill-intentioned small reef fish mimic the behavior of the cleaners so well that the host receives them readily; however, instead of cleaning parasites, the intruders nibble the unsuspecting host's flesh.

Our flashlight illuminates a tapestry of bright color in the reef environment, much of it contributed by an extraordinary variety of reef fish. Let's pause motionless over a patch of reef and watch the resident animals feed. Nearly every kind of feeding behavior may be found on a reef. Small crabs and shrimp eat detritus. Coral polyps and the giant clam *Tridacna*, which may achieve a length of 1 meter, are filter-feeders, straining minute plankton from surrounding waters. Some animals, like moray eels, nourish themselves by preying directly on other animals. Other predators, including the reef-destroying crown-of-thorns starfish *Acanthaster planci*, nibble the living polyps. Parrot fish bite off whole chunks of coral with their strong toothy jaws, grind the calcareous skeletons with their pharyngeal dentition (teethlike structures located in the throat), and obtain their nutrition from the abundant filamentous algae embedded in the coral.

The embedded filamentous algae are macroscopic and distinct from the microscopic, unicellular zooxanthellae which inhabit the

TOP VIEWS

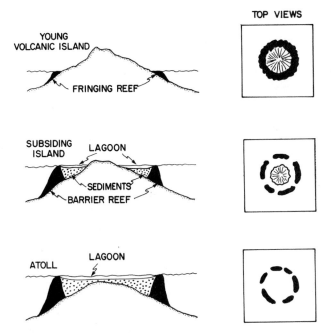

FIGURE 11.29. Illustration of Darwin's theory of atoll formation. (Adapted from a diagram by Nelson Marshall and reproduced by permission.)

living tissues of the coral. Though they are not readily conspicuous because they grow within the coral skeletons, they are generally the most important and abundant primary producers on the reef. They make up nearly 75 percent of the organic material in coral and 94 percent of the photosynthetic tissue. A careful perusal of the reef surface should give you an indication of why these algae are so important. Large fleshy algae are conspicuously absent. Although they will grow well in this environment, heavy grazing by reef herbivores, including sea urchins and algae-eating fish, prevent them from getting larger than stubble. In contrast, embedded filamentous algae are protected from most grazers.

There are exceptions to the general lack of large seaweeds on a coral reef. The few species that do thrive are relatively unpalatable. These include the *coralline algae*, which possess an exoskeleton of calcium carbonate and which provide more calcareous material than do corals to some reefs.

Several different types of coral reefs are recognized. Since these very often represent stages in the development of a reef, it is instructive to examine the processes of reef formation and growth, first hypothesized by Charles Darwin and confirmed by scientists a century later. Coral atoll formation is depicted in Figure 11.29.

Most tropical islands are of volcanic origin. After cooling and solidifying, their great weight causes them to gradually sink back down into the earth's mantle through subsidence (see Chapter 3). Initially, the corals which grow in the shallow waters around the island's perimeter form a *fringing reef*. As the island sinks, carrying the old coral down with it, new coral grows on top of the old so that the upper surface of the reef remains in well-illuminated waters. As the island continues to sink, the land mass grows increasingly distant from the encircling coral; eventually, a *barrier reef* is formed, with a lagoon between the reef and the island. Finally, the entire island sinks below the sea surface, leaving a still-healthy ring of coral with an enclosed lagoon. Such a reef is called an *atoll* and is exemplified by Enewetak Atoll in the Pacific Ocean.

SUGGESTED READINGS

BARNES, R. S. K., AND HUGHES, R. N. 1982. *An introduction to marine ecology.* Chapters 3–6. Oxford: Blackwell.

FOX, W. T. 1983. *At the sea's edge: an introduction to coastal oceanography for the amateur naturalist.* Englewood Cliffs, N.J.: Prentice-Hall.

MACLEISH, W. H. 1976. Estuaries. *Oceanus* 19(5). (A collection of articles.)

NYBAKKEN, J. W. 1982. *Marine biology, an ecological approach.* Chapters 5–10. New York: Harper and Row.

PHILLIPS, R. C. 1978. Seagrasses and the coastal marine environment. *Oceanus* 21:30-40.

SUMICH, J. L. 1984. *An introduction to the biology of marine life.* 3rd ed. Chapters 3–4. Dubuque, Iowa: Brown.

TAIT, R. V. 1981. *Elements of marine ecology.* 3rd ed. Chapter 8. London: Butterworths.

Chapter Twelve

Ocean Destinations

DREAM A LITTLE! DREAM OF A GLOBAL, SOLO VOYAGE IN A SMALL but seaworthy sailboat. Dream of discovering for yourself, recording, and interpreting the major features of the global ocean.

For romantic (and practical) reasons your vessel, *Ocean Quest*, is a replica of the *Spray*, the sturdy, 37-foot rebuilt oyster sloop (later yawl) used by Joshua Slocum in his pioneering, single-handed circumnavigation of the globe from 1895 to 1898. You have installed a vane for automatic steering and a small diesel engine, primarily to keep your batteries charged. You will use relatively simple scientific instruments which you've been introduced to in a comprehensive oceanography course, and a few reference texts which your professor has suggested. You'll have food and water for up to sixty days, replenishing at ports along your route. You'll supplement your diet with fish.

You plan to use prevailing winds and currents whenever you can, and you'll allow yourself some time in most of the major oceans and seas (Figure 12.1).

You leave in late spring, departing from Newport, Rhode Island. Off Nantucket lightship, you head toward Cape Race, Newfoundland. The prevailing westerlies are blowing, and with the wind astern you make good time. Fortunately, the weather is good! You're in one of the Atlantic's busiest ocean highways, and you don't want to be run down by the big boys.

You're going to try to record weather, sea conditions, and surface-water temperature and salinity (with a refractometer) every four hours. Once a day, you'll use Nisken bottles to check deeper-water temperature and salinity every 10 meters to a depth of 50 meters. You'll also make daily tows for phytoplankton and zooplankton. You'll try to identify and quantify the plankton with the aid of your field microscope and reference texts.

You have a lot of water to sail over. The Atlantic covers 82 million square kilometers and averages over 3 kilometers deep. Though only half the area of the Pacific, one of these days the Atlantic may actually be larger because, due to plate tectonics (see Chapter 4), it is currently expanding (by a few centimeters a year) while the Pacific is contracting.

At first your plankton tows bring an abundance of organisms, dominated by diatoms and copepods. This is because you are over the continental shelf. Its shallow, well-mixed waters receive considerable nutrients from land, deeper water, and the sea floor and are well illuminated in late spring (see Chapter 9). Mixing is even

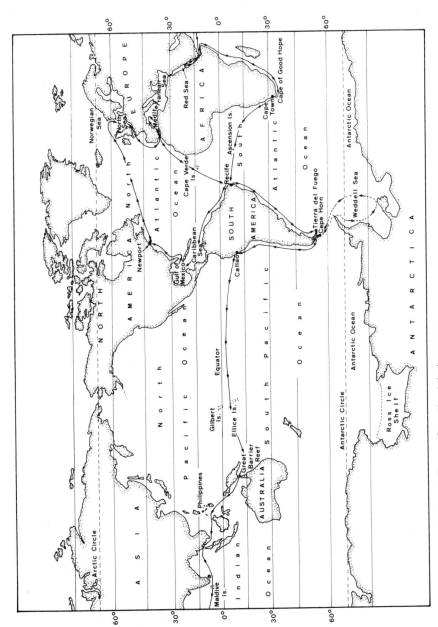

FIGURE 12.1. Route of *Ocean Quest*'s global voyage.

greater at this time of year because the summer thermocline isn't yet established. Though phytoplankton are numerous, the spring "bloom" that typifies many temperate latitude waters actually peaked weeks ago and has since been grazed down by zooplankton. You also find abundant fish larvae: Many North Atlantic species spawn in the spring.

The fertile conditions of the North Atlantic shelf waters create some of the richest fisheries in the world, particularly off the northeast United States and Canadian Maritime Provinces, and in the North, Barents, and Norwegian Seas. The Atlantic has the largest catch per unit area of any ocean and accounts for nearly as large a harvest as the Pacific.

A drop in surface-water temperature indicates that you may now be bucking the Labrador Current, the sweep of cold water which originates between the west coast of Greenland and Baffin Bay, flows southward into the North Atlantic, then swings around Newfoundland toward the Gulf of Maine.

Fog! This is a persistent feature of these waters, caused by the encounter between the Labrador Current and the comparatively warm westerlies. And maybe icebergs! They are progeny of the mile-thick Greenland icecap; some 12,000 are calved every year as the glaciers flow into the sea. They are first swept northward by the West Greenland Current and then southward into the Atlantic by the Labrador Current. They typically weigh over a million tons and tower over 60 meters above the sea surface (and several hundred meters below the surface). They're insulated by the cold sea water which surrounds most of their volume, but they gradually shrink to about 10 percent of their original size by the time they reach the Atlantic.

Finally, the fog disappears. The air is warmer. Squall clouds appear off to the southwest. The previously murky green waters are now a deep blue. Water temperatures have risen several degrees. You refer to the Gulf Stream plot which you obtained from the Woods Hole Oceanographic Institution just prior to your departure. You're right in a northward-looping meander of the sinuous current. Such meanders frequently curl back on themselves, creating large independent eddies which spin away from the Stream and persist on their own for several weeks (see Chapter 7).

Soon you're being aided by the North Atlantic Current. This derives from the Gulf Stream and forms the northern portion of the North Atlantic gyre. It is the warm current which softens the climate of western Europe. Broader than the Gulf Stream, it flows at only about one-tenth the speed of the latter, averaging about a quarter of a knot. On reaching the eastern side of the ocean basin, part of the North Atlantic Current branches into a westward-flowing loop, the Irminger Current, which eventually joins the southward-flowing Greenland Current to form the counterclockwise Subpolar gyre. The remainder divides into the northward-flowing Norwegian Current (which you'll follow) and the southward-flowing Canary Current, which comprises the eastern portion of the North Atlantic gyre.

Soon your navigational fixes find you just west of the British Isles. You sail carefully through Scotland's offshore islands and enter the southern fringes of the Norwegian Sea. The summer weather has augmented the fishing fleet which works these fertile waters. Most boats are trawling for cod. This important North Atlantic species comprises several different stocks, each with its own distinct spawning areas, feeding grounds, and migration routes. Like haddock, flounder, ocean perch, and hake, cod are bottom-dwelling (*demersal*), feeding on small crustaceans, marine worms, and the larvae of other fish. The females spawn between January and March, laying several million eggs which take 10 to 20 days to hatch.

After a brief layover in Bergen, Norway, you enjoy exploring Norwegian fjords before entering the North Sea. You pass several more trawlers fishing for cod, haddock, and plaice. You also sail by a purse seiner sweeping the sea for herring. These species all spawn in the region's fertile waters. The North Sea is less than 1 percent the size of the Atlantic, but it has historically been one of the world's richest fishing grounds. Its productivity is largely attributable to its shallow depths—about 100 meters on average. This allows continuous mixing of its nutrient-rich waters, promoting a large production of plankton.

Geologically, the North Sea is a young water body; 250 million years ago this area was a desert plain surrounded by mountains! Since then it has periodically flooded and dried out with the retreat and advance of ice ages. In the process, over 6 kilometers of sediments have accumulated. These are now the basis for an important sand and gravel industry.

You draw abeam of an oil-production platform and marvel at the immensity of this triple-decked structure which supports 100 workers, can drill over 2 dozen wells, and is designed to withstand winds of over 100 knots and wave heights of nearly 30 meters. The

North Sea's thick marine shales and porous sandstones act as sources and traps for oil and gas and are responsible for the extensive offshore petroleum industry in the region today.

Sailing southward, enroute to the Mediterranean, you transit the heavily traveled English Channel. You pass the hulk of an old wreck which alerts you to the dangers of the long, unstable sand bars created when the Channel's strong currents continually move the loose sea-floor sediments. The numerous, shifting wrecks which have resulted are themselves a navigational hazard.

You arrive at Gibraltar in early fall and enter the Mediterranean. This nearly landlocked sea is about 3 million square kilometers in area. It is approximately five times the size of the North Sea, but it would take over thirty Mediterraneans to cover as much area as the Atlantic. Its average depth is 1500 meters, less than half that of the Atlantic. It is about 4000 kilometers long and is essentially divided into two basins by a narrow constriction at the Strait of Sicily.

These two basins are quite different. If the sea floor were visible beneath *Ocean Quest*, you would see a major east-west–oriented ridge. The Mediterranean Ridge System bisects the eastern basin. Though it's superficially similar to the Mid-Atlantic Ridge, it is not a spreading center but rather the result of compression of sediments in the basin, a process thought to be related to Alpine mountain building in the region. In contrast, the geologically younger western basin doesn't have a major ridge and is characterized by rather smooth abyssal plains and two large basins.

Active sea-floor movements throughout the Mediterranean account for the region's well-known volcanic activity. Mts. Etna and Stromboli are among many volcanoes that are active today.

Deep drilling studies in the Mediterranean recently revealed a surprising finding: The presence of thick layers of evaporites indicates that the Mediterranean has completely dried out a number of times during its history, turning the seabed into an arid desert landscape.

The complex circulation of the Mediterranean is markedly different from that of most of the world's major oceans (see Chapter 7). It's in an arid location in the Horse Latitudes. Its depths are shallow. These factors, combined with its landlocked nature and limited exchange of water with the rest of the world's oceans, mean that water loss through evaporation is about three times as great as its replacement from runoff and precipitation.

This creates an interesting phenomenon. In compensation for evaporative water loss from the Mediterranean, Atlantic Ocean water constantly flows in through the Strait of Gibraltar. Less saline (and less dense) than the Mediterranean water, which has been made saltier by evaporation, the entering Atlantic water establishes a surface current which carries on through the west and east basins. It is the only well-defined current in the Mediterranean. As it flows eastward, its salinity increases through evaporation and the surface waters begin to sink. They sink still further with the onset of winter as the lowered temperature raises their density even more. Then the cold, salty bottom waters gradually flow back along the bottom to the Strait of Gibraltar where they spill over a sill at the entrance and sink down into the Atlantic. Mediterranean water is readily apparent in the Atlantic because of its comparatively high salinity; it can be detected well out into the center of the ocean to a depth of nearly 1000 meters!

Your tows show much lower plankton densities than you found in the North Atlantic. You expected this and know that the Mediterranean's unique circulation is largely responsible. Surface water entering at Gibraltar has been stripped of nutrients by Atlantic phytoplankton. Runoff and decomposition introduce some nutrients, but most of these sink, are trapped by the stable water column, and are carried back to the Atlantic by the return flow along the bottom. Low nutrient concentrations mean low productivity.

You are not prepared, however, for a raft of dead fish which you sail through near a large coastal city. You wonder if pollution is responsible. The Mediterranean is particularly vulnerable to pollution. Because its exchange of waters with the Atlantic is so restricted, waste products from surrounding land areas tend to accumulate in the basins. Discharged sewage is one of the worst problems. Most of it is untreated, and its decomposition severely lowers oxygen concentrations in some areas. Some 300,000 tons of oil are also introduced to the Mediterranean every year, primarily from oil tankers legally discharging waste in "free zones." Pollution has badly hurt the region's fishing industry.

Surprisingly, the Mediterranean fishery, though less productive than many others, is potentially the world's most valuable. This is because the resource is not only accessible but also brings a high price. Unfortunately, this high price has resulted in severe overharvesting of some stocks.

After your minicruise into the "Med," you head southwest for another transoceanic passage, this time across the South Atlantic toward Brazil, riding the northeast trades.

The Canary Current gives you an extra boost. That's the portion of the North Atlantic gyre which flows southwest off the coast of North Africa. It's a broad, fairly weak current which will assist you nearly halfway to Brazil.

The wind dies completely, then begins to freshen. Soon a moderate gale is in progress. Panic! This is hurricane season in the tropical Atlantic. But the weather forecasts and a careful watch of the barometer indicate no major nearby depression. Thank goodness, you have simply encountered one of the frequent gales characteristic of this region throughout the year.

A period of fine weather follows. On a sparkling day, a cloud bank looms ahead on an otherwise clear horizon. You are approaching the Cape Verde Islands. The cloud bank is caused by condensation of warm, moist air rising off the land mass, which meets the cooler air over the surrounding ocean.

Just north of the equator, a period of intermittent calms, fitful squalls and thunderstorms, and a slackening of the trades indicates that you are now getting into the unpredictable doldrums (see Chapter 6). Here the northeast and southeast trade winds converge, and high evaporation rates and subsequent condensation of moist air aloft cause considerable tropical storminess. You also find that you're in the Equatorial Countercurrent, which flows eastward at about 1 knot just north of the equator.

So you head south across the equator and celebrate the event with a tepid bottle of champagne which you have slightly cooled by lowering it to 50 meters below while becalmed in the doldrums.

By now you've tangled with several of the Atlantic current systems. One hundred meters below you still another major current silently glides by. This is the Equatorial Undercurrent, which flows eastward at about 2 knots. Some 200 kilometers wide, its flow is seldom shallow enough to affect shipping, but the energy in its upper layers has been known to hinder the steering of fully loaded supertankers.

In the calms, you have fun seeing the marine life that has been attracted to the food and shelter provided by the now-luxuriant growth on the boat's bottom and the strange, floating shape that is *Ocean Quest*. Besides small fish and invertebrates, which seem to have taken up residence below the waterline, swordfish, yellowtails, dolphins, sharks, and porpoises frequently follow your boat for days. Morning usually finds your boat littered with flying fish, just in time for breakfast.

Plankton tows indicate high biological productivity here at the equator. Since nutrients are usually scarce in tropical surface waters because renewal is blocked by a permanent thermocline (see Chapter 9), productivity there is usually low. The fertility here is attributable to equatorial upwelling, which brings deep, nutrient-rich water to the surface (see Chapter 7). You also record the lowest salinity since you left Gibraltar. This is because of high rainfall at the equator.

Early one late-fall morning you hear the cries of sea birds. This gives a hint of land not far off. Not long after you raise the coast of Brazil and dock in Recife for supplies and repairs. You also purchase a small cabin stove. You're going to need it on the next leg of your journey.

A few days later you're off again. You're hoping to catch up with the National Science Foundation polar research vessel, *Hero*, because you've wangled a berth on her through the influence of your oceanography professor. She's going to spend the southern hemisphere summer exploring Antarctic waters.

Your progress toward Tierra del Fuego, near Cape Horn, is boosted by the Brazil Current, which is the southward-flowing western boundary current of the South Atlantic gyre. This is the equivalent of the Gulf Stream but is a much slower current, usually averaging less than 2 knots. It also has a much smaller volume than the Gulf Stream because it represents only a portion of its precursor, the South Equatorial Current, which divides off the coast of Brazil. The other portion flows northwest as the Guiana Current.

Just beyond Cape San Antonio at the southern extremity of the Rio de la Plata estuary, your progress is slowed despite a fair, following wind. You are now bucking the Falklands Current. This is an important cold-water current which works its way inside the Brazil Current and surges north to about 30 degrees south latitude.

South of Bahia Blanca you are forced to ride out a series of westerly gales as you sail down the coast of Argentina. Finally, the weather abates near Cape Horn. Sea birds become numerous. Most are shearwaters or members of several petrel species. There are numerous whalebirds, whose pale bluish and white coloration camouflages them nicely in the rough seas. Your first albatross! It's a large white bird with black on its upper wing surfaces and an extraordi-

nary wingspan of over 3 meters. This is an adult wandering albatross, a common bird of the Southern Ocean. It spends three quarters of the year at sea. You admire the way it soars. It seldom beats
its wings but adjusts them for wind speed and direction, effortlessly
riding the air currents much like a sailplane. It's joined by others,
looking for discarded garbage, a bonanza they have come to expect
from vessels in these waters.

As you sail past the entrance to the Strait of Magellan, intermittent cold rain and persistent fogs make for unpleasant sailing,
summer though it is. Rough seas continue. Water temperature drops
to less than 10° C (only about 5 degrees warmer than in midwinter).
Your discomfort is momentarily forgotten when you catch a
glimpse of the south coast of Tierra del Fuego. The desolate, barren
rocks typify this part of the world. High mountains and snow-cov-
ered peaks are visible through the gloom. Close to shore you make
out penguins diving off ledges and frolicking in the whitecaps. Like
porpoises, these aquatic birds often leap well clear of the water
while they swim at speeds as high as 20 knots.

On you sail through the Beagle Channel, a narrow passage
thought to have had its origins in a rift created about 150 million
years ago when South America, Africa, and Antarctica split apart
from what had formerly been the supercontinent, Pangaea (see
Chapter 4).

You've made it just in time to join *Hero*, docked in Ushuaia,
Tierra del Fuego. She's a wooden two-masted sailboat, 125 feet
long, and carries an experienced crew of thirteen. She's been designed for Antarctic conditions, with a low center of gravity and a
rugged tapered hull which resists crushing by ice by rising upwards
when pressure is increased at the waterline. You'll earn your keep
by assisting with hydrographic observations and plankton counts.

Underway in *Hero*, you soon cross the Antarctic Convergence,
which essentially marks the northern boundary of the Antarctic
Ocean. Here, northward-flowing Antarctic Intermediate Water
sinks below the warmer and lighter waters of the Subantarctic
Zone, north of about 55 degrees south latitude. A vertical profile at
the Convergence shows the several distinct water masses sand-
wiched between the surface and the bottom which are described in
Chapter 7.

The "surface" currents of the Antarctic are primarily established by the prevailing winds but are also influenced by density
differences between water masses. The westerly gales of the "Roar-

ing Forties" (the perpetually stormy latitudes of the Southern Ocean below about 40 degrees south latitude) are the driving force behind a major eastward-flowing circumglobal current, the Antarctic Circumpolar Current (also called the West Wind Drift; see Chapter 7). Though it moves slowly, with speeds averaging less than a quarter of a knot, it transports more water than any other ocean current. In contrast to the shallow currents of warmer, thermally layered ocean areas, the nonstratified Antarctic Circumpolar Current runs deep—its flow may be detected to a depth of several kilometers! South of 60 degrees south latitude, the surface circulation flows toward the west. That current is called the East Wind Drift because it is influenced by prevailing easterly winds there.

You're excited to finally be in the Antarctic, the earth's fourth largest ocean. It comprises 35 million square kilometers, slightly less than half the size of the Atlantic. Sixty percent of it is covered by sea ice in the winter, and 4 million square kilometers are permanently frozen. The ocean is effectively divided into three distinct basins—the Indian, Pacific, and Atlantic—by sections of the global oceanic ridge system on its northern border. The floor of these basins lies nearly 5 kilometers deep.

After several days of sailing, you sight mountainous Elephant Island, located in the South Shetland archipelago. This is the island that provided temporary shelter for heroic explorer Ernest Shackleton and his beleaguered crew after their ship was destroyed by ice in 1915.

The results of hydrocasts and plankton tows are somewhat surprising. In open, offshore areas the abundance of plankton and large animals, while substantial, is smaller than you expected, despite ample nutrients and well-mixed waters. Productivity is clearly higher near land masses and areas of melting ice. This is probably related to the amount of mixing in surface waters. In the open ocean, fertile conditions may be partly wasted because the sinking of water masses and generally rough conditions can cause many phytoplankton to be mixed down below the compensation depth (see Chapter 9). In contrast, the reduced salinity and density of surface waters near land and melting ice may allow phytoplankton to remain in the euphotic zone longer, stimulating productivity throughout the food chains.

A day beyond Elephant Island, you sight the Antarctic continent. Larger than the United States and Mexico combined, it has a total area of almost 9 million square kilometers. Surprisingly, it is the highest continent on earth, averaging 1800 meters above sea level. Ninety-eight percent of the continent is blanketed by ice and snow. Nine-tenths of the earth's total ice and snow is here, nearly 5 kilometers thick in places! The Polar Plateau, the huge glacier which covers the continent's interior, has a major effect on the entire world's weather. In fact, climatology is a primary interest of the thirty-five Antarctic research stations maintained by eleven countries.

Much of Antarctica's ice is located in vast ice shelves which extend far into the sea from the edges of the continent. The Ross Ice Shelf alone has an area nearly as large as that of France.

Life is abundant here, and all of it is dependent on the ocean. There is a distinct absence of land mammals in Antarctica. Marine birds are especially prevalent. They include penguins, Cape pigeons, cormorants, terns, gull-like fulmars, predatory skuas, and the omnipresent petrels and albatrosses. The endurance of the Arctic terns is extraordinary. Though only half the size of a herring gull, they migrate over 30,000 kilometers every year, flying from the Antarctic to breed in the Arctic. Bare rocky areas inevitably harbor rookeries of Adelie, gentoo, or chinstrap penguins. Their pebble nests are packed so closely together that it is virtually impossible to walk through a colony without stepping on eggs or young chicks.

Marine mammals, including seals, sea lions, and whales are much in evidence. You see a pod of humpback whales straining planktonic krill (see Chapter 9) through the filterlike baleen structures which hang from the roofs of their mouths. Like many of the birds, some whale species, including the humpback, migrate thousands of miles to and from their breeding grounds to feed in the productive waters of the Antarctic. Tragically, commercial whaling has reduced global whale populations to perhaps 10 percent of former levels.

Giant kelp is abundant on the rocky shores of islands north of 60 degrees south latitude. This is *Macrocystis pyrifera*, the common, valuable marine plant of cold, nutrient-rich waters. Among the largest (as long as 55 meters) and fastest-growing (up to a third of a meter a day) plants in the world, giant kelp is harvested heavily off the west coasts of North and South America for its alginic acid. High costs of collection and processing currently prohibit commercial exploitation in the Antarctic.

Summer has nearly ended, and it's time to head back to Ushuaia. Already, recently unfrozen areas are beginning to ice over.

At oceanic salinities, the freezing temperature of water is lowered to −1.9° C (see Chapter 5). At that point, water molecules assume the latticelike matrix characteristic of ice crystals. Close packing of the crystals and sea-surface turbulence result in the eventual formation of thin ice plates with raised edges; these are collectively referred to as *pancake ice*. In continuing cold, pancake ice coalesces and thickens into an uninterrupted ice floe. But because it insulates the water below, the thickness of the floe is limited so that, over a span of time, wind stress fractures the floe, creating narrow leads, wider pools called *polynyas*, and large pressure ridges wherever moving masses of ice meet and pile up.

Idly, you munch on a small piece of floating ice, noting that it tastes surprisingly fresh because the salt is excluded from ice crystals during freezing. At first some of the salt is trapped in miniscule brine cells within the ice, but eventually that too escapes. That's why older ice is virtually salt-free.

Back in Ushuaia, you reboard *Ocean Quest* and head northwest. Interesting as your trip to the Antarctic has been, you're glad to be underway again in your own boat. Finally, you're in the Pacific, largest of all oceans, so massive that it covers one third of the globe—more than all land areas combined! It's also the deepest ocean, with an average depth of over 4 kilometers and a maximum depth, in the Marianas Trench, of 11 kilometers.

As you sail along the coastline of Chile, you can make out the majestic Andes mountain range, thrusting up to 6.5 kilometers above sea level. Directly below you, the Peru-Chile Trench plunges to an equivalent depth. The relief difference between the land and immediately adjacent sea floor is the greatest on earth—a direct ascent from the bottom of the trench to the mountain peaks would require a vertical climb of some 13 kilometers over a horizontal distance of slightly more than 300 kilometers. That's steep!

The physical proximity of the Peru-Chile Trench and the Andes mountain range is not coincidental. The oceanic ridge which cleaves the Pacific Basin is considerably displaced to the east as the East Pacific Rise. This oceanic spreading center separates the massive Pacific Plate, which moves slowly away to the northwest, from the smaller Nazca Plate, which slides away to the southeast until it abuts the westward-moving American Plate, carrying the continent of South America at its leading edge. Where they meet, the Nazca Plate grinds below the American Plate, in the process forging the Peru-Chile Trench. It's the frictional contact between plates above this subduction zone that results in the volcanic activity that has created the Andes mountain range.

Again you get an assist from a current, this time the north-bound Peru Current. This is also called the Humboldt Current. It comprises the eastern boundary of the South Pacific gyre and flows north at a speed of about one third of a knot (only about 10 percent of the velocity of the Atlantic's Gulf Stream). Just south of the equator the current is divided into a coastal and an oceanic component by the southward-flowing Peru Countercurrent. The Countercurrent seldom ventures below 2 to 3 degrees south latitude, but underlying all of the surface currents the broad Peru Undercurrent surges slowly southward.

Your water samples show that the Peru Current is cold. Between 50 degrees and 35 degrees south latitude, the temperature has warmed only from 11° C to 15° C. Even near the equator the temperature of the Peru Current seldom rises above 22° C. This keeps the west coast of South America much cooler than regions of equivalent latitude on the east coast.

Why are the waters of the Peru Current so cold? In part it is due to the influence of the Antarctic Circumpolar Current, but mostly it is attributable to coastal upwelling, described in Chapter 7. As the southeast trade winds, somewhat skewed by the Andes, blow parallel to the coasts of Peru and northern Chile, the near-shore surface waters move out to sea (through Ekman Transport; see Chapter 7) because of the deflecting influence of the Coriolis effect. In response, deep, cold waters rise to the surface, replacing the volume driven away. You can't see this upwelling—no telltale turbulence disturbs the sea surface. It is a very slow and gradual process. The water rises vertically less than 100 meters a month.

Your samples turn up a lot of plankton. You are not surprised. The upwelled waters contain substantial nutrients, which stimulate productivity. Besides, cold water contains more of the vital gases (including carbon dioxide, required for photosynthesis) than warm water.

One morning, cormorants, boobies, and pelicans are plunging into the water all around you, feeding on a swarm of small fish. One fish, dropped on *Ocean Quest*'s deck, is a Peruvian anchovy, a member of the herring family about 10 centimeters long. This may be the most important fish in the world. It thrives in the nutrient-rich waters of the Peru-Chile upwelling zone and feeds on abundant minute animals, which in turn are fueled by vast fields of phytoplankton. The anchovies are commercially harvested for conversion to fish meal and fish oil and constitute a major fishery. In the early 1970s so many anchovies were harvested that Peru led the world in fish production—with nearly a quarter of the globe's catch!

The birds that alerted you are part of the greatest population of sea birds in the world. Feeding principally on the anchovies, their guano has accumulated over the years to depths of nearly 50 meters on some offshore islands! An excellent natural fertilizer, its recovery is a profitable industry.

This enormous productivity of Peru's coastal waters can't always be taken for granted. Sometimes the upwelling ceases, and the harvest falters. Just as the winds trigger the upwelling, so are they responsible for its failure. Unpredictably, the southeast trades periodically slacken or are supplanted by westerly breezes. Without the driving force of the trade winds, the Peru Current weakens and gives way to the Countercurrent whose warm, relatively low-salinity waters spread inexorably southward. Soon a layer of low-density, nutrient-poor waters, 30 meters thick, blankets the coastal zone. This blocks upwelling and initiates a catastrophic decline in marine productivity, link by link through the food chains. The disastrous warm water intrusion, popularly known as El Niño, is now known to be coupled to a series of meteorological events of global significance termed the El Niño/Southern Oscillation (see Chapter 6).

El Niño is not the only culprit responsible for the virtual collapse of the Peruvian anchovy fishery since the early 1970s; years of severe overharvesting have also taken their toll and impeded the fishery's natural recovery.

In mid-May you're in the port of Callao, Peru, stocking supplies and preparing for the longest leg of your journey so far—the 16,000-kilometer trek across the South Pacific. It was here at Callao that *Kon-Tiki* was launched and started out on her trans-Pacific voyage.

You ride the tail end of the Peru Current as it curves away from the coast. As the days pass, the current flow increases and the waves mount higher. But still the waters are cold. Finally, the seas moderate, the temperatures rise, and the water color has again gone from coastal green to the clear blue of less productive offshore tropical seas. You have entered the South Equatorial Current, a major component of the South Pacific gyre. Driven by the southeast trades, it flows westward at about 1 knot between 3 degrees and 10 degrees south latitude.

This is not the only major current that runs across the Pacific here. Below the limit of your subsurface sampling gear, a thin, narrow undercurrent flows eastward toward Peru. This is the Cromwell Current, the best-known undercurrent in the world's oceans. It moves at a speed of about 3 knots, from 70 to 200 meters below the sea surface.

Caught in the great sweep of the South Equatorial Current and propelled forward by the steady breath of the southeast trades, you make good progress across the Pacific. It's a period of quiet days and restful nights. It's a time for reflection and relaxation. But it is also a time for science.

Without the rude intrusion of changing weather and storms, this period is what your dreams were made of. Life is very much in evidence. Dolphins and tunnies are your constant companions. Sometimes there is the uneasy company of surfacing whales. And sharks! Always with a vanguard of pilot fish patrolling inches from their hosts' snouts.

Some creatures share environments from water to air and back again. The graceful aerial excursions of flying fish and the shorter, jet-propelled flights of small squid can fascinate you for hours.

There is even more activity at night. The phosphorescent glow of tiny bioluminescent plankton sparkles in your wake. On calm nights, large, vague shapes loom beneath the hull, hinting of mysteries in the depths. Nighttime samples confirm diurnal vertical migration (see Chapter 10); surface tows show how zooplankton abundance increases from dusk to midnight and then declines precipitously after dawn. After dark, bizarre deep-sea fish come to surface waters, perhaps following prey fish migrating above them.

But the results of your plankton tows don't indicate high fertility here. Although there is an impressive variety of species in your fine-meshed nets (and it would be even more impressive if your sampling gear could capture the tinier plankton, which are even more abundant), the *density* of phytoplankton and zooplankton is the lowest you have encountered thus far on your long journey. These warm, well-illuminated waters are relatively unproductive because a permanent thermocline stratifies the water column and prevents mixing of accumulated deep nutrients into the upper layers of the sea.

The low concentrations of nutrients in tropical surface waters are somewhat offset by year-round high light intensities, enhanced by the waters' sparkling clarity. Plankton production throughout most of the tropics is slow but steady, in marked contrast to the boom-or-bust cycles of higher-latitude waters. Sampling year-round shows almost half as much annual production here as in fertile temperate areas.

Your speed slows perceptibly as you begin to buck the eastward-flowing South Equatorial Countercurrent. This is better developed in the Pacific than in the Atlantic, with two components on either side of the equator, separated by a branch of the Equatorial Current, which flows in the opposite direction.

You alter your course slightly to the southwest to put yourself back in the South Equatorial Current. This takes you through the chain of volcanic islands which makes up the Gilbert-Ellice group. You know from your reading that, like the Hawaiian chain, these islands originate from a "hot spot" in the Pacific Plate (see Chapter 4). What you cannot see is that the island chain is oriented north-south, with the oldest islands to the north. In contrast, the islands of Hawaii stretch from southeast to northwest, where the older geological formations are, reflecting the present direction of plate movement. The Gilbert-Ellice Islands are much older than the Hawaiian Islands. During their explosive birth, more than 40 million years ago, the Pacific Plate was moving in a northerly direction.

After an idyllic cruise throughout the heart of the Pacific islands of everyone's dreams, you sail on toward the Australian continent, still borne along by the reliable South Equatorial Current. The sweep of its waters has now veered toward the south, enroute to a new identity as the East Australian Current, the relatively weak western boundary current of the South Pacific gyre. But along the way it caresses Australia's outer shelf and deposits you by one of the greatest oceanographic wonders of the world—the Great Barrier Reef.

Despite all the background reading, you're hardly prepared for your first view of the Great Barrier Reef. Bathed in the spreading glow of the rising tropical sun, it's an endless necklace of foaming waters and sparkling islets that curves away to the north and south. It's the world's largest reef area. Its verdant islands and lush coral gardens form a band, 24 to 290 kilometers wide, which rims the entire coast of northeast Australia—some 1600 kilometers altogether.

The reef's rough waters discourage a close approach, but the constant aeration provided by the eternal surf is essential for its prolific growth. Tidal range on the reef averages 3 to 4 meters and exceeds 10 meters in some locales. The resulting strong tidal currents funnel through reef openings, sweeping in vital materials and flushing out wastes. From May to November the southeast trades drive the waters onto the reef; during the other months, the summer monsoon whistles down from the north.

But productive reefs need more than moving water. The water must also be warm—coral reefs do not develop where water temperatures average less than 18° C. The tepid South Equatorial Current keeps temperatures on the reef at 27° C.

Coral reefs also need substantial light to permit photosynthesis by the coral polyps' symbiotic algae, the zooxanthellae (see Chapter 11), and by the larger algae which provide much of a reef's structure. The Great Barrier Reef's shallow depths (only 30 to 55 meters) allow light to penetrate to the bottom. The clarity of its waters also aids light penetration and coral growth. Reefs do not develop successfully in turbid, sediment-laden waters, no matter how warm and well nourished.

Like other coral reefs, the Great Barrier requires abundant dissolved nitrogen and phosphorus to nurture its resident algae. Though tropical waters are usually deficient in these nutrients, there are ample supplies here. Upwelling is partly responsible. As two branches of the South Equatorial Current diverge in the Coral Sea, deep, nutrient-rich waters rise up to fill the void created by the parting surface waters.

Productive coral reefs also need high concentrations of dissolved carbonate ions so that the coral animals and calcareous algae can build the external skeletons which make up the reef's limestone substratum. Again, upwelling of carbonate-rich deep water in the Coral Sea supplies this need.

Skirting the northeast perimeter of the Great Barrier Reef with help from a branch of the South Equatorial Current that meanders northwest, you sail through the narrow Torres Strait and along the southwest coast of New Guinea. Then you're in Indonesia's myriad islands. Finally, in mid-March, you enter the Celebes Sea, enroute to the Philippines' Sulu Archipelago.

Although the Celebes Sea has a well-deserved reputation for piracy, you cross it safely and sail over the shallow reef flats of Sulu's 300-odd islands. Here the islanders are utterly dependent on the sea for sustenance, income, and transportation.

The extent of this dependence is apparent as you approach a village built entirely over the water, its thatched houses perching precariously on pilings driven into the shallow bottom. The village is built that way to free up scarce land for cultivation, to facilitate waste disposal, and to defend against marauders.

Further along you encounter a cluster of small outrigger canoes which house entire families of sea gypsies. These *bajaos*, as they are locally known, refuse to set foot on land, presuming dire consequences if they leave their watery habitat. A small palm roof is their only shelter; within this tiny, cramped microcosm of life, the full cycle of human events proceeds, often accompanied by the full measure of human suffering.

A kilometer beyond the bajao community there is a network of upright stakes. They are made of resilient mangrove wood and support rows of nylon monofilament line from which large clumps of brownish-green algae are suspended. This is a seaweed farm, one of dozens tended by island families in these waters.

The crop cultivated here is *Eucheuma*, a marine plant in great demand by several chemical companies because its viscous extract, carrageenan, is a valuable gelling and stabilizing agent for various foods and pharmaceuticals. Dwindling supplies, resulting from overharvesting of wild stocks, paved the way for the introduction of its cultivation in the early 1970s.

Cultivation of seaweeds accounts for a sixth of the world's production from aquaculture—the farming of freshwater and marine organisms. Fish account for most of the rest—at least 40 percent of these are grown in freshwater lakes in China!

Aquaculture is one of the oldest professions on earth, but despite glowing predictions about its future contribution to world food supplies it is, so far, well entrenched in only a dozen or so countries. These are primarily in Asia and produce 90 percent of all cultivated aquatic crops. Because cultivation allows close control over an organism's environment, nutrition, reproduction, and even genetic composition, technological advances can vastly improve crop productivity and quality.

You leave Sulu and work down the north coast of Sabah, Malaysia. Your next leg takes you through the Straits of Malacca into the Bay of Bengal in the eastern Indian Ocean.

After the Pacific and Atlantic, the Indian Ocean is the world's third largest body of water. It makes up about one fifth of the globe's total water area. Its average depth is nearly 4 kilometers, and its deepest extremity, the Java Trench, lies 7.5 kilometers below the sea surface.

Your late fall arrival gives you a westward passage during the northern hemisphere winter, when the winds and currents of the prevailing northeast monsoon are favorable (see Chapter 7 for a description of the Indian Ocean's seasonal monsoons and accompanying surface circulation). With the winds off your starboard quarter and the seasonal North Equatorial Current carrying you along, you head toward the Red Sea, over 5000 kilometers to the west.

You make a landfall in mid-December. It is the Maldives, a cluster of volcanic islands which comprises an archipelago about 1000 kilometers southwest of Sri Lanka.

As you approach, you come on the white water of a coral reef, the seaward reef margin of the immense Suvadiva Atoll, whose length of 70 kilometers and width of nearly 60 kilometers make it one of the world's largest atolls. The waters here seem calmer than those of the Great Barrier Reef, and you welcome the chance to explore the reef and inner lagoon.

Although you can't see it from the boat, a "spur-and-groove" system of alternating coral ridges and channels extends seaward at right angles to the reef itself. These structures effectively baffle incoming waves, acting as natural breakwaters which absorb up to 95 percent of the waves' energy.

You locate a quiet channel, or *pass*, on the leeward side of the reef which provides access to the lagoon from the open ocean. The bottom is clearly visible. The middle of the pass is largely devoid of corals, probably because of the scouring effect of sand swept in and out with the tides. On either side of the pass entrance, the encircling seaward *reef margin* consists of a slightly elevated ridge of coral colonies and encrusting calcareous algae (a pronounced *algal ridge* occurs on the reef's windward side). These algae may be as important as corals in providing structure to some reefs. Just inside the reef margin lies a flat expanse of living corals, dead coral fragments, and patches of sand. This *reef flat*, as it is called, is partially exposed at low tide. It is considerably narrower than its counterpart on the windward side of the atoll, since the water is more fertile and often less turbid on the more exposed side, allowing faster growth of the reef.

At the inside boundary of the reef flat a platform of rock slopes gradually to the enclosed, protected lagoon. As you scan the lagoon, you notice that it is rimmed by several small islets scattered around the inside edge of the reef flat. Most harbor groves of coconut trees and other terrestrial vegetation which can withstand frequent exposure to salt spray, drought, and abrasion by windblown sand.

Inside the lagoon there are numerous isolated clusters of corals, called *patch reefs*, clearly visible in the clear water. You can tell their approximate depth from the color of the water as it shifts from the blue to the red end of the spectrum with decreasing depth. The average depth of water inside the lagoon is about 25 meters, but it varies considerably and generally increases toward the center. The lagoon water is slightly more turbid than that in the pass because of accumulated sediments, mostly deriving from coral debris and decaying vegetation. The lagoon's shallow, confined waters are also warmer than those over or outside the reef and undergo more salinity fluctuations in response to rainfall and high evaporation rates.

The mechanism of atoll formation was first hypothesized by Charles Darwin in 1842 and subsequently confirmed by test drillings after World War II (see Chapter 11). An atoll is a circle of small islands and submerged reefs which enclose a central lagoon. Its development starts after the formation of a volcanic island in tropical waters. Corals grow at sea level on the flanks of the island and create a fringing reef. Gradually, the island's concentrated weight causes it to sink back into the earth's mantle, but the upward growth of corals and algae keeps pace with this slow subsidence and forms a barrier reef and lagoon. Eventually, through subsidence and erosion, the volcanic island is completely submerged. The overlying barrier reef and lagoon constitute an atoll.

After anchoring, you go snorkeling. The potpourri of marine life in the lagoon is fascinating and includes a large variety of coral forms, brightly colored reef fish, moray eels, spiny sea urchins, gastropods, brittle stars, and starfish (see Chapter 11 for a more detailed description of coral reef life). You're tempted to stay and stay, but you must move on.

As you clear the Maldives, you heave to and watch a small fleet of outrigger sailing canoes gather in a net full of fish. Fishing methods here have changed little over the centuries, and fishing is still primarily a small-scale subsistence operation. Although the water area of the Indian Ocean is nearly nine tenths that of the Atlantic, its fish catch is less than one fifth that of the Atlantic. Primitive techniques don't make for large catches!

But further along you pass a fleet of large, sleek Taiwanese tuna vessels, a reminder that modern fishing techniques are being introduced to this ocean. Harvests have gone up rapidly in recent years: Between 1965 and 1974, total catch increased by 50 percent!

In early January, you enter the Gulf of Aden and proceed through the Bab el Mandeb to the Red Sea. This warm, salty sea is an embryonic ocean which, given time, could become one of the largest on the surface of the globe. How could this be?

The secret lies about 2000 meters below you. Running from the Gulf of Suez to the Gulf of Aden, a long rift bisects the sea floor. This deep axial trough and valley is, in fact, an oceanic ridge and spreading center, with the volcanic activity and patterns of sediment distribution and magnetic orientation characteristic of all such spreading centers (see Chapter 4). The spreading appears to have continued uninterrupted for the past 2 million years; as the Arabian Plate moves away to the northeast and the Africa Plate slides to the southwest, the Red Sea widens by about a centimeter a year—a rate which, in 200 million years, would make it as wide as the Atlantic!

You head back out into the Indian Ocean and sail down the east coast of Africa. Outside the Gulf of Aden a tongue of warm, salty water flows out of the Red Sea and sinks below the surface of the Indian Ocean, to a depth of nearly a kilometer.

In the Mozambique Channel, between Africa and the island of Madagascar, you get a boost from the warm Agulhas Current, the western boundary current of the South Indian gyre. With speeds often approaching 4 knots, this is the strongest western boundary current south of the equator. It achieves its maximum flow during the northeast monsoon, when reinforced by waters of the South Equatorial Current. But enormous waves sometimes result when the swift current encounters large swells racing northward from Southern Ocean storms. And you're also concerned about the weather which lies ahead in the Roaring Forties off the Cape of Good Hope.

Fortunately, the summer weather prevails for your rounding of the notoriously dangerous Cape. Only slightly worse for the wear, you dock in Cape Town, where you prepare for your third trans-Atlantic crossing and the final stages of your long trip.

Underway again, you head northwest in the slow drift of the Benguela Current, the cold-water current which makes up the eastern portion of the South Atlantic gyre. You continue to ride the currents and follow the sweep of the gyre back toward Recife, altering your course progressively to the west as the Benguela Current is succeeeded by the broad swath of the South Equatorial Current. After the anxious anticipation off the Cape of Good Hope, you welcome the comparative languor of these uneventful days and content yourself with observations of marine life, your sampling routine, and reading.

Your route takes you just south of Ascension Island, a tiny pinnacle of volcanic rock on the mid-Atlantic Ridge. Just west of Ascension, you see a green sea turtle (genus *Chelonia*), an ocean-going reptile which can achieve weights of up to 100 kilograms. What is it doing so far from the mainland?

In one of the most extraordinary feats of animal navigation yet documented, a population of green turtles migrates regularly from feeding grounds in Brazil to Ascension Island, 2200 kilometers away in the mid-Atlantic. There they mate, nest, and lay their eggs. How they unerringly home in on a tiny midocean land mass, only 8 kilometers wide, and do so by swimming against the South Equatorial Current is still largely a mystery.

By mid-May you are in Brazil, renewing acquaintances in Recife. On the last leg of your odyssey, you sail up the northeast coast of South America and loop up through the Caribbean. Your speed is boosted by about 2 knots by the northwest-flowing Guiana Current. Like the Brazil Current, this warm current (temperatures of 26–28° C) derives from the South Equatorial Current off the coast of Brazil.

In mid-June you're off the island of Grenada, and you log in your entry to the Caribbean Sea. This semienclosed water body is roughly the size of the Mediterranean and is surrounded by the mainland of South and Central America to the west and south and by the islands of the Lesser Antilles, Puerto Rico, Dominican Republic/Haiti, and Cuba to the east and north. These islands and connecting ridges restrict the exchange of deep water with the Atlantic Ocean and account for a much smaller tidal range in the Caribbean and Gulf of Mexico than in the Atlantic.

Although the Caribbean has been extensively studied, its geological history is still uncertain. Its floor consists of the small, almost stationary Caribbean Plate, which may be an ancient fragment of the Pacific Plate. As larger plates slide past or drive into the Caribbean Plate, the surrounding earth groans and shudders. In fact, most of the pronounced volcanic and earthquake activity around the Caribbean basin is caused by the Pacific and North American Plates plunging below the Caribbean Plate.

You sail west to the northern coast of South America and detour into Lake Maracaibo, a semienclosed embayment in northwestern Venezuela. The number of offshore drilling platforms, associated with the region's extensive oil fields, is astounding. During the peak demand years of the early 1970s, this area produced nearly 4 million barrels of oil per day, making Venezuela the world's third largest oil-producing country!

As you head northwest toward the Gulf of Mexico, you pass a small fleet of boats seining for anchovy. This is the first truly mechanized fishing operation you have seen since entering these waters. The Caribbean has a relatively unproductive fishery because of limited accessible continental shelf areas, low levels of nutrients in surface waters, and the absence of major unwellings. There are more modern and productive fisheries in the Gulf of Mexico, where large harvests of shrimp and menhaden are made.

Passing through the Yucatan Channel, you enter the Gulf of Mexico, borne along by the Caribbean Loop Current which makes a sharp turn around Cuba, funnels through the Straits of Florida, and joins the Gulf Stream east of Miami.

With an area of about 1.5 million square kilometers and an average depth of about 1500 meters, the size and depth of the Gulf of Mexico are equivalent to the dimensions of the Mediterranean. Like the Mediterranean, the Gulf of Mexico is also semienclosed. The surrounding land masses and underwater ridges help create diurnal tides (see Chapter 8) and a small tidal range in the Gulf, in contrast to the adjacent Atlantic Ocean which experiences larger, semidiurnal tides.

As you draw near New Orleans and the U.S. coast, you again sail by a number of oil and gas drilling rigs. You must be over the continental shelf. The Gulf of Mexico's extensive shelf areas average 100 to 240 kilometers in width and are overlain by a thick blanket of sediments which harbor large quantities of petroleum hydrocarbons. As you approach the Louisiana coast, the increasingly turbid water signals the major source of the northern Gulf's sediments.

Every year, the Mississippi River collects some 360 million tons of eroded materials from America's heartland and pours them into the Gulf of Mexico! This tremendous sediment load has created a major delta seaward of New Orleans which extends all the way across the continental shelf. Your chart clearly shows this delta and the network of small branching channels which feeds it. As the finer sediments are carried further seaward, they are either deposited on the steep slope at the front of the delta or are swept westward by prevailing longshore currents.

As you head southeast, you rejoin the Caribbean Loop Current and approach the Straits of Florida. It is now late July, and you are anxious to head north on the final leg of your journey, before hurricane season begins in earnest. South of Miami you enter the Florida Current, the southernmost component of the Gulf Stream system. Swept along by the Florida Current and, north of Cape Hatteras, by its successor, the Gulf Stream, your speed is boosted by up to 3 or 4 knots, and you make excellent time up the Atlantic seaboard.

On a muggy morning in late August you are off the Brenton Reef tower, outside Narragansett Bay, Rhode Island. Not long after, you dock at the same berth from which you sailed over two years ago. As you secure *Ocean Quest* and mentally prepare for the transition back to life ashore, you reflect on your long journey. In slightly less than 28 months you have sailed through every ocean except the Arctic (and you came close to that). You've visited most of the major seas. You have lived your dream. The oceans of the world are real! But you still have a long way to go. There is much to digest and much more to learn. In many ways your dream voyage has just begun!

Appendixes

APPENDIX 1. Metric Table and Conversion Factors

LENGTH

1 kilometer (km) = 1000 (10^3) meters = 0.621 statute mile = 0.540 nautical mile
1 meter (m) = 100 (10^2) centimeters = 39.4 inches = 3.28 feet = 1.09 yards
1 centimeter (cm) = 10 millimeters = 0.394 inch
1 millimeter (mm) = 1000 (10^3) microns = .0394 inch
1 micron (μ) = one thousandth (10^{-3}) mm = .0000394 inch
1 statute mile = 1609 meters = 5280 feet
1 nautical mile = 1852 meters = 6076 feet = 1.151 statute miles
1 fathom = 1.8288 meters = 6 feet

AREA

1 square kilometer (km^2) = 1,000,000 square meters = 0.386 square statute mile = 0.292 square nautical mile
1 square meter (m^2) = 10,000 square centimeters = 10.76 square feet
1 square centimeter (cm^2) = 100 square millimeters = 0.155 square inch

VOLUME

1 cubic meter (m^3) = 1,000,000 cubic centimeters = 1000 liters = 35.3 cubic feet
1 cubic centimeter (cm^3) = 0.061 cubic inch
1 liter = 1000 cubic centimeters = 1.06 quarts

MASS

1 gram (g) = 0.035 ounce
1 kilogram (kg) = 1000 grams = 2.205 pounds
1 metric ton = 1,000,000 grams = 1000 kilograms = 2205 pounds

TEMPERATURE

Absolute zero = 0° Kelvin (K) = −273.2° Celsius (C) = −459.7° Farenheit (F)
Fresh water freezes at 273.2° K = 0.0° C = 32.0° F
Conversions: K = °C + 273.2

$$C = \frac{(°F - 32)}{1.8}$$

$$F = (1.8 \times °C) + 32$$

SPEED

1 meter per second (m/s) = 100 cm/s = 3.281 ft/s = 1.944 knots
1 kilometer per hour (km/hr) = 0.277 m/s = 0.909 ft/s = 0.55 knot
1 knot = 0.51 m/s = 1.151 statute miles per hour = 1 nautical mile per hour

Kingdom Monera: single-celled; lacking membrane-bound organelles such as a nucleus; exemplified by blue-green algae and bacteria.

Kingdom Protista: single-celled and having nuclei and other membrane-bound organelles; exemplified by diatoms, dinoflagellates, silicoflagellates, coccolithophores, and marine protozoans, including radiolarians, foraminiferans, and ciliates.

Kingdom Plantae: photosynthetic and multicellular organisms.

> **Phylum Chlorophyta:** the "green" seaweeds.

> **Phylum Phaeophyta:** the "brown" seaweeds, including economically and ecologically important kelps and rockweeds.

> **Phylum Rhodophyta:** the "red" seaweeds, including many which are economically important.

> **Phylum Tracheophyta:** having vascular tissue and true leaves, stems, and roots.

>> **Class Angiospermae:** flowering plants which produce seeds inside fruit; include the salt marsh grasses of the genus *Spartina*, mangrove trees, and submerged eelgrass, turtlegrass, and widgeon grass of the genera *Zostera*, *Thalassia*, and *Ruppia*.

Kingdom Animalia: multicellular heterotrophic organisms whose principal means of nourishment is the ingestion of other organisms; existing as plankton, nekton, or benthos in the oceans.

> **Phylum Porifera:** the sponges.

> **Phylum Cnidaria (Coelenterata):** radially symmetrical with stinging cells and tentacles; include jellyfish, anemones, and corals.

> **Phylum Ctenophora:** comb jellies; transparent planktonic animals which superficially resemble jellyfish but move by means of comb plates of fused cilia.

> **Phylum Platyhelminthes:** flatworms which include free-living forms and parasitic flukes and tapeworms of fish and other marine animals.

> **Phylum Nemertina:** ribbon worms; similar to flatworms but more advanced, elongated, and possessing a retractable proboscis.

> **Phylum Aschelminthes:** roundworms; lacking segmentation, usually small, and including the widespread and abundant nematodes.

> **Phylum Chaeotognatha:** arrowworms; torpedo-shaped, planktonic, predatory animals usually less than 3 centimeters long.

Phylum Mollusca: mollusks; often (but not always) with one or more shells; include clams, mussels, snails, and squids.

Phylum Annelida: segmented worms; elongate and lacking a rigid skeleton.

> *Class Polychaeta:* marine annelids which include tube-dwelling forms (tube worms) and free-living forms (notably the familiar bait worms).

Phylum Arthropoda: having a hard, jointed outer skeleton with paired, jointed appendages.

> *Class Chelicerata:* horseshoe crabs.

> *Class Crustacea:* includes copepods, crabs, shrimp, lobsters, and barnacles.

Phylum Echinodermata: radially symmetrical animals which are exclusively marine and possess a spiny skin and a system of water-filled canals (the water vascular system) which aids in locomotion and food capture.

> *Class Crinoidea:* feather stars or sea lilies.

> *Class Asteroidea:* starfish.

> *Class Echinoidea:* sea urchins.

> *Class Holothuroidea:* sea cucumbers.

Phylum Hemichordata: acorn worms.

Phylum Chordata: animals which, at some stage in their development, have a notochord (a flexible, internal supporting rod), a dorsal, hollow nerve cord, and gill slits in the pharynx region.

> *Subphylum Urochordata:* tunicates; animals in which the larval stage resembles a tadpole but the adult is sedentary and sac-like.

> *Subphylum Vertebrata:* animals having a backbone (made of cartilage or bone) and a cranium (brain case).

> > *Class Agnatha:* jawless fish; include lampreys.

> > *Class Chondrichthyes:* fish having jaws and cartilaginous skeletons; include sharks and rays.

> > *Class Osteichthyes:* fish having jaws and bony skeletons; include herrings, salmon, flatfish, tuna, and cods.

> > *Class Reptilia:* air-breathing animals with dry, scaly skin whose young develop in eggs; include sea turtles and sea snakes.

> > *Class Mammalia:* warm-blooded animals (homeotherms) with milk-producing glands and, often, body hair and whose young are born alive; include whales, seals, sea lions, walruses, sea otters, manatees, and dugongs.

> > *Class Aves:* warm-blooded, feather-bearing animals whose forelegs have been modified as wings; include sea birds.

Index